T0338491

WIND RESOURCE
ASSESSMENT

WIND RESOURCE ASSESSMENT

A Practical Guide to Developing a Wind Project

Michael C. Brower

AWS Truepower, LLC, Albany, New York, USA

with contributions from

Bruce H. Bailey
Philippe Beaucage
Daniel W. Bernadett
James Doane
Matthew J. Eberhard
Kurt V. Elsholz
Matthew V. Filippelli
Erik Hale
Michael J. Markus
Dan Ryan
Mark A. Taylor
Jeremy C. Tensen

A JOHN WILEY & SONS, INC., PUBLICATION

Library of Congress Cataloging-in-Publication Data:

Wind resource assessment : a practical guide to developing a wind project / Michael C. Brower... [et al.].
      p. cm.
   Summary: "This is a practical, authoritative guide for the most important phase in developing a wind
energy project"—Provided by publisher.
   ISBN 978-1-118-02232-0 (hardback)
  1. Wind power. 2. Wind power plants. I. Brower, Michael, 1960-
   TJ825.W555 2011
   621.31'2136–dc23

                                                                                   2011040044

10 9 8 7 6 5 4 3 2 1

The authors would like to thank the New York State Energy Research and Development Authority (NYSERDA) for supporting the creation of the Wind Resource Assessment Handbook (Albany, New York, USA; NYSERDA; 2010), on which parts of the present book are based. We also gratefully acknowledge the help of the following reviewers, who provided many useful comments and corrections to the manuscript: Dennis L. Elliott, Matthew Hendrickson, Ian Locker, Kathleen E. Moore, Ron Nierenberg, Andrew Oliver, Gordon Randall, Marc Schwartz, and Richard L. Simon. Notwithstanding their diligent efforts, any errors and oversights remain the sole responsibility of the authors. We would also like to thank the following staff of AWS Truepower for their assistance in preparing the manuscript: Alicia Jacobs, Alison Shang, and Amber Trendell. Their help is deeply appreciated.

Michael C. Brower (ed.)
Bruce H. Bailey
Philippe Beaucage
Daniel W. Bernadett
James Doane
Matthew J. Eberhard
Kurt V. Elsholz
Matthew V. Filippelli
Erik Hale
Michael J. Markus
Dan Ryan
Mark A. Taylor
Jeremy C. Tensen

CONTENTS

PREFACE

I was a young physicist just out of graduate school when I first became familiar with wind energy. At the time (the early 1990s), the industry was in its infancy. Although wind turbines had been deployed by the thousands, they were mostly small machines, seemingly prone to breaking down and collectively supplying only a tiny fraction of the world's electricity needs.

How far we have come! Today's utility-scale wind turbines are enormous structures half as high as a supertanker's length. Soaring gracefully over the landscape, they generate up to several megawatts of power each and run more reliably and quietly than a car. The industry, too, has advanced by leaps and bounds. By the end of 2010, wind accounted for nearly 4% of the world's generating capacity, and it will likely overtake both nuclear power and hydroelectric power before this decade is over (http://www.gwec.net/fileadmin/images/Publications/GWEC_annual_market _update_2010_-_2nd_edition_April_2011.pdf).

Every industry, as it matures, needs standards. They provide assurance that the industry's products will perform as advertised, which helps draw the consumers and investors it needs to prosper. This is especially true of wind resource assessment—the process by which wind power developers and consultants estimate how much electricity a plant will produce. A mistake at this critical stage of a project's development can mean disastrous losses for its financial backers, as well as a black eye for the wind industry.

There is no question that plenty of mistakes have been made in the past. Resource assessment methods, including everything from anemometer calibration and mounting standards to the modeling of terrain and vegetation effects, have evolved as the industry has learned what works and what doesn't. As a consequence, many projects have not performed as well as expected.

Fortunately, that period of experimentation is mostly behind us, and the industry has coalesced around a reasonably consistent set of practices and standards that provide good confidence that a plant's expected production will be realized. Communicating that body of knowledge to practitioners of wind resource assessment is the key aim of this book.

To be sure, not every aspect of resource assessment is fully mature. There are continuing debates over topics such as remote sensing and numerical wind flow modeling. What's more, new challenges are constantly emerging as wind turbines and projects grow in size and more advanced methods of measuring and modeling the wind are introduced. While this book cannot definitively settle these debates or anticipate every innovation, we hope that readers will gain enough information and insight to make sound decisions about the tools and methods they should use.

This book is a collaborative effort by a team of experts in wind resource assessment, all of whom were employed at the time of writing by AWS Truepower, LLC, a renewable energy consultancy based in Albany, New York. Different authors wrote and contributed to different chapters, which were then edited into a coherent whole. The result, we hope, is a practical, authoritative guidebook that will serve the industry well for years to come.

MICHAEL C. BROWER

Albany, NY, USA

1

INTRODUCTION

For any power plant to generate electricity, it needs fuel. For a wind power plant, that fuel is the wind.

Wind resource assessment is the process of estimating how much fuel will be available for a wind power plant over the course of its useful life. This process is the single most important step for determining how much energy the plant will produce, and ultimately how much money it will earn for its owners. For a wind project to be successful, accurate wind resource assessment is therefore essential.

Technologies for measuring wind speeds have been available for centuries. The cup anemometer—the most commonly used type for wind resource assessment—was developed in the mid nineteenth century, and its basic design (three or four cups attached to a vertical, rotating axis) has scarcely changed since.

Yet, an accurate estimate of the energy production of a large wind project depends on much more than being able to measure the wind speed at a particular time and place. The requirement is to characterize atmospheric conditions at the wind project site over a wide range of spatial and temporal scales—from meters to kilometers and from seconds to years. This entails a blend of techniques from the mundane to the sophisticated, honed through years of experience into a rigorous process.

Wind Resource Assessment: A Practical Guide to Developing a Wind Project, First Edition.
Michael Brower et al.
© 2012 John Wiley & Sons, Inc. Published 2012 by John Wiley & Sons, Inc.

The details of this process are the subject of this book. Before diving into them, however, we should back up a little and set wind resource assessment in context. Where does the wind come from? What are its key characteristics? And how is it converted to electricity in a wind power plant?

1.1 WHERE DO WINDS COME FROM?

The simple answer to this question is that air moves in response to pressure differences, or gradients, between different parts of the earth's surface. An air mass tends to move toward a zone of low pressure and away from a zone of high pressure. Left alone, the resulting wind would eventually equalize the pressure difference and die away.

The reason air pressure gradients never completely disappear is that they are continually being powered by the uneven solar heating of the earth's surface. When the surface heats up, the air above it expands and rises, and the pressure drops. When there is surface cooling, the opposite process occurs, and the pressure rises. Owing to differences in the amount of solar radiation received and retained at different points on the earth's surface, variations in surface temperature and pressure, large and small, are continually being created. Thus, there is always wind somewhere on the planet.

While uneven solar heating is ultimately the wind's driving force, the earth's rotation also plays a key role. The Coriolis effect[1] causes air moving toward the poles to veer to the east, while air heading for the equator veers to the west. Its influence means that the wind never moves directly toward a zone of low pressure but rather, at heights above the influence of the earth's surface, it circles around it along the lines of constant pressure. This is the origin of the cyclonic winds in hurricanes.

By far, the most important temperature gradient driving global wind patterns is that between the equator and the poles. Combined with the Coriolis effect, it is responsible for the well-known equatorial trade winds and midlatitude westerlies (Fig. 1-1). At the equator, relatively warm, moist air has a tendency to rise through convection to a high altitude. This draws air in from middle latitudes toward the equator and thereby sets up a circulation known as a *Hadley cell* (after the nineteenth century meteorologist who first explained the phenomenon). Because of the Coriolis effect, the inflowing air turns toward the west, creating the easterly trade winds.[2]

A similar circulation pattern known as a *polar cell* is set up between high latitudes and the poles. Lying between the polar and Hadley cells are the midlatitude (Ferrel) cells, which circulate in the opposite direction. Unlike the others, they are not driven by convection but rather by the action of sinking and rising air from the adjacent cells. Once again the Coriolis effect asserts itself as the air flowing poleward along

---

[1] The Coriolis effect is a property of observing motions from a rotating reference frame—in this case, the earth. The earth's surface moves faster around the axis at the equator than it does closer to the poles. If an object moves freely toward the equator, the surface beneath it speeds up toward the east. From the perspective of an observer on the surface, the object appears to turn toward the west.

[2] By convention, wind direction is denoted by the direction the wind comes *from*. If the air is moving toward the north, it is said to be a southerly wind.

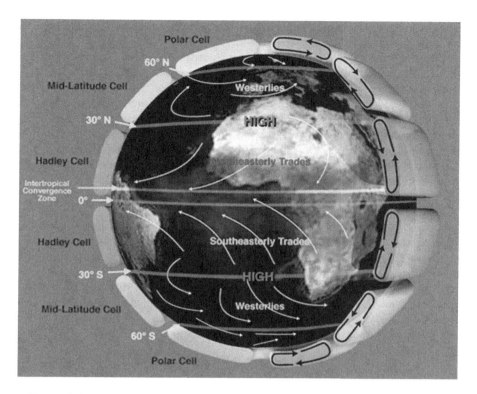

<u>Figure 1-1.</u> The main global atmospheric circulations. *Source:* NASA/JPL-Caltech.

the surface turns east, creating the westerlies. The westerlies are the reason wind resources tend to be so good in the temperate and high latitudes (around 35–65 °N) of North America, Europe, and Asia, as well as the southern extremes of Africa, South America, and Australia.

Superimposed on these global circulation patterns are many regional patterns. Large land masses heat up and cool down more rapidly than the oceans, and even within land masses, there are variations in surface heating, for instance, between a snow-covered mountain top and a green valley below or between a desert and a cultivated plain. The resulting temperature gradients set up what are called *mesoscale atmospheric circulations* — mesoscale because they are in between the global scale and the local scale, or microscale.

The most familiar mesoscale circulation is the sea breeze. During a typical summer day, the land becomes warmer than the ocean, the pressure drops as the air above it expands and rises, and relatively cool, dense air is pulled in from the ocean. At night, the process reverses, resulting in a land breeze. Normally, sea breezes are weak, but where the wind is concentrated by terrain, they can have a powerful effect. This is the primary mechanism behind the very strong winds found in coastal mountain passes in the US states of California, Oregon, and Washington, and in comparable passes in other countries.

While temperature and pressure differences create the wind, it can be strongly influenced by topography and land surface conditions as well, as the example of coastal mountain passes attests. Where the wind is driven over a rise in the terrain, and especially over a ridge that lies transverse to the flow, there can be a significant acceleration, as the air mass is "squeezed" through a more restricted vertical space. Thanks to this effect, many of the best wind sites in the world are on elevated hilltops, ridges, mesas, and other terrain features. However, where the air near the surface tends to be cooler and heavier than the air it is displacing, as in the sea breeze example, it has a tendency to find paths around the high ground rather than over it. In such situations, it is often the mountain passes rather than the mountain tops that have the best wind resource.

Surface vegetation and other elements of land cover, such as houses and other structures, also play an important role. This role is often represented in meteorology by a parameter called the *surface roughness length*, or simply the roughness. Because of the friction, or drag, exerted on the lower atmosphere, wind speeds near the ground tend to be lower in areas of higher roughness. This is one of the main reasons why the eastern United States has fewer good wind sites than, for example, the Great Plains. Conversely, the relatively low roughness of open water helps explain why wind resources generally improve with distance offshore.

1.2   KEY CHARACTERISTICS OF THE WIND

The annual average wind speed is often mentioned as a way to rate or rank wind project sites, and indeed, it can be a convenient metric. These days, most wind project development takes place at sites with a mean wind speed at the hub height of the turbine of 6.5 m/s or greater, although in regions with relatively high prices of competing power or other favorable market conditions, sites with a lower wind resource may be viable. However, the mean speed is only a rough measure of the wind resource. To provide the basis for an accurate estimate of energy production, the wind resource must also be characterized by the variations in speed and direction, as well as air density, in time and space.

1.2.1   The Temporal Dimension

The very short timescales of seconds and less is the domain of turbulence, the general term for rapid fluctuations in wind speed and direction caused by passing pressure disturbances, or eddies, which we typically experience as brief wind gusts and lulls. Turbulence is a critical mechanism by which the atmosphere gradually sheds the energy built up by solar radiation. Unfortunately, it has little positive role in power production because wind turbines cannot respond fast enough to the speed variations. In fact, high turbulence can cause a decrease in power output as the turbine finds itself with the wrong pitch setting or not pointing directly into the wind. In addition, turbulence contributes to wear in mechanical components such as pitch actuators and yaw motors. For this reason, manufacturers may not warrant their turbines at sites where the turbulence exceeds the design range. Knowledge of turbulence at a site is thus very important for resource assessment.

Fluctuations in wind speed and direction also occur over periods of minutes to hours. Unlike true turbulence, however, these variations are readily captured by wind turbines, resulting in changes in output. This is a time frame of great interest for electric power system operators, who must respond to the wind fluctuations with corresponding changes in the output of other plants on their systems to maintain steady power delivery to their customers. It is consequently a focus of short-term wind energy forecasting.

On a timescale of 12–24 h, we see variations associated with the daily pattern of solar heating and radiative cooling of the earth's surface. Depending on the height above ground and the nature of the wind climate, wind speeds at a given location typically peak either at midafternoon or at night. Which pattern predominates can have a significant impact on plant revenues in markets that price power according to the demand or time of day. For example, regions in which air-conditioning loads are important often see a peak in power demand in the afternoon, and regions in which there is heavy use of electricity for home heating may experience a peak in the early evening.

The influence of the seasons begins at timescales of months. In most midlatitude regions, the better winds usually occur from late fall to spring, while the summer is less windy. Sites experiencing strong warm-weather mesoscale circulations, such as the coastal mountain passes mentioned earlier, are often an exception to this rule, and winds there tend to be strongest from late spring to early fall. Because of seasonal variations like this, it is difficult to get an accurate fix on the mean wind resource with a measurement campaign spanning much less than a full year. Furthermore, as with diurnal variations, seasonal variations can impact plant revenues. Power prices are usually the highest in summer on a summer-peaking system and in winter on a winter-peaking system.

At annual and longer timescales, we enter the domain of regional, hemispheric, and global climate oscillations, such as the famous El Niño. These oscillations, as well as chaotic processes, account for much of the variability in wind climate from year to year. They are the main reason why it is usually desirable to correct wind measurements taken at a site to the long-term historical norm.

1.2.2 The Spatial Dimension

The spatial dimension of wind resource assessment is especially important for wind plant design. Most wind power plants have more than one wind turbine. To predict the total power production, it is necessary to understand how the wind resource varies among the turbines. This is especially challenging in complex, mountainous terrain, where topographic influences are strong. One approach is to measure the wind at numerous locations within the wind project area. Even then, it is usually necessary to extrapolate the observed wind resource to other locations using some kind of model, typically a numerical wind flow model.

The spatial scales of interest are related to the size of wind turbines and the dimensions of wind power projects. The rotors of modern, large wind turbines range in diameter from 70 to 120 m. Wind turbines are typically spaced some 200–800 m apart, and large wind projects can span a region as wide as 10–30 km. Within this overall range, a detailed map of the variations is essential for the optimal placement of wind turbines and accurate estimates of their energy production.

The vertical dimension is just as important. The variation in speed with height is known as *wind shear*. In most places, the shear is positive, meaning the speed increases with increasing height because of the declining influence of surface drag. Knowing the shear is important for projecting wind speed measurements from one height (such as the top of a mast) to another (such as the hub height of a turbine). Extreme wind shear (either positive or negative) can cause extra wear and tear on turbine components as well as losses in energy production. The shear is typically measured either by taking simultaneous speed readings at more than one height on a mast or with a remote sensing device such as a sodar (sonic detection and ranging) or lidar (light detection and ranging).

1.2.3   Other Characteristics of the Wind Resource

Although wind speed is the dominant characteristic of the wind resource, there are other important ones, including wind direction, air density, and icing frequency, all of which need to be well characterized to produce an accurate energy production estimate.

Knowledge of the frequency distribution of wind directions is key for optimizing the layout of wind turbines. To reduce wake interference between them (described below), turbines are generally spaced farther apart along the predominant wind directions than along other directions.

Air density determines the amount of energy available in the wind at a particular wind speed: the greater the density, the more energy is available and the more electric power a turbine can produce. Air density depends mainly on temperature and elevation.

A substantial amount of ice accumulating on turbine blades can significantly reduce power production, as it disrupts the carefully designed blade airfoil, and can become so severe that turbines must be shut down. The two main mechanisms of ice accumulation are freezing precipitation and direct deposition (rime ice). Other conditions potentially affecting turbine performance include dust, soil, and insects.

1.3   WIND POWER PLANTS

Conceptually, a wind turbine is a simple machine (Fig. 1-2). The motion of the air is converted by the blades (lifting airfoils very similar to airplane wings) to torque on a shaft. The torque turns a power generator, and the power flows to the grid.

However, this simple picture disguises many subtle design features. The typical modern large wind turbine is an immense, complicated machine ranging from

Figure 1-2. Utility-scale wind turbines. *Source:* AWS Truepower.

65 to 100 m in height at the hub, with a rotor 70–120 m in diameter, and with a rated capacity of 1–5 MW. The turbine must operate reliably and at peak efficiency under a wide range of wind conditions. This requires numerous components, from nacelle anemometers to pitch actuators and yaw drives to power electronics, working together in an integrated system.

Perhaps, the key characteristic of a wind turbine from the perspective of wind resource assessment is the turbine power curve (Fig. 1-3). This describes the power output as a function of wind speed measured at the hub. It is characterized by a cut-in speed, typically around 3 or 4 m/s, where the turbine begins turning and generating power; a sloping portion, where the output increases rapidly with speed; a rated speed, typically around 13–15 m/s, where the turbine reaches its rated capacity; and a cut-out speed, above which the turbine control software shuts the turbine down for its protection.

Although well-operated turbines are finely tuned machines, it is wrong to assume that a turbine produces exactly the expected power at every wind speed. For example, blade wear and soiling, equipment wear, and control software settings can all cause turbines to deviate from their ideal power curve. In addition, power output depends on wind conditions, such as turbulence, the variation of wind speed across the rotor, and the inclination of the wind flow relative to horizontal. Taking account of such variations is part of the process of estimating energy production, and it starts with a detailed understanding of the wind resource.

Wind power plants are likewise conceptually simple: they are just arrays of wind turbines linked through a power collection system to the power grid (Fig. 1-4). However, designing a wind project often entails delicate trade-offs between, for example, total plant output and construction cost.

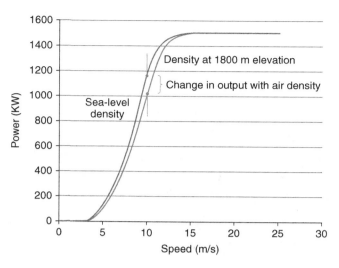

Figure 1-3. Typical power curve for a 1.5-MW turbine at two different air densities. *Source:* AWS Truepower.

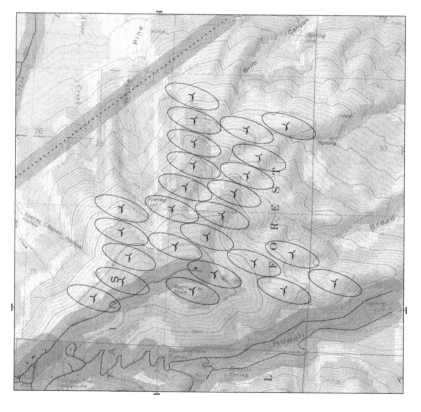

Figure 1-4. Layout of a proposed wind farm. *Source:* AWS Truepower.

Figure 1-5. Rare visual evidence of turbine wakes in an offshore wind farm. The increased turbulence behind each turbine causes the water vapor in the air to condense as droplets, forming a visible contrail. The wind speed in each wake is also reduced. *Source:* Horns Rev 1 owned by Vattenfall. Photographer Christian Steiness.

The process begins by producing a detailed picture of how the wind resource is distributed across the site, supported by measurements and spatial modeling of some kind. In sites with complex terrain and wide variations in land cover, this can be a significant technical challenge. A further complication is wake (or array) interference between turbines. When a turbine extracts energy from the wind, a zone of reduced wind speed and increased turbulence is created behind it (Fig. 1-5). Any turbines that happen to be within this wake will generally produce less power than if the upwind turbines were not there. Fortunately, wakes tend to expand and dissipate with distance downwind as turbulence exchanges energy with the surrounding, undisturbed wind flow. How the wakes from all the turbines impact plant production is usually estimated with a specialized wake model.

1.4 PURPOSE AND ORGANIZATION OF THIS BOOK

As we have seen, designing a wind project and estimating its energy production depend on a detailed and accurate assessment of the wind resource, which is where this book comes in. The book is primarily intended to give guidance to practitioners and students on the accepted methods of wind resource assessment for utility-scale wind farms. The goal is not to impose conformity in every respect. On the contrary, the book often highlights areas where there is room for reasonable variation, even disagreement, on the approaches that can be used. Nonetheless, the range of variation has limits. It may be acceptable in some cases to install a tower with just two levels of anemometers, rarely just one. It may be fine to use a new or unusual atmospheric model, but not without anchoring the results in reliable measurements or testing the model's accuracy. What we hope the reader gains from this book is a clear understanding of those limits.

Whenever possible, the book goes beyond a cookbook to describe some of the concepts and principles behind the tried-and-true techniques. This, we hope, will empower the reader to make his or her own judgments where conditions depart, as they often do, from the ideal. What the book does *not* strive to be is a comprehensive reference on every aspect of wind resource assessment. For those many interesting topics, there are standards published by the International Electrotechnical Commission (IEC), proceedings of the many wind conferences that occur every year around the world, and a number of books and Internet-based resources.

The book is organized in the order of the main stages of wind resource assessment. The first several chapters focus on the nuts and bolts of conducting a wind measurement campaign. It starts with an overview of the wind resource assessment process. Then it moves through site selection, measurement parameters and tower instrumentation, tower installation and maintenance, and data collection and handling. The last chapter in this group, Chapter 8, focuses on remote sensing (lidar and sodar).

The next section of chapters addresses how the wind resource data are analyzed. It starts with quality control (QC) and validation and then moves on to characterizing the observed wind resource. Subsequent chapters cover extrapolating the resource estimates to hub height, correcting short-term measurements to long-term historical conditions, and wind flow modeling. Chapter 14 is devoted to special issues concerning offshore sites. Chapter 15 discusses uncertainty in wind resource estimates, including the different categories of uncertainty and their typical values. The last chapter, Chapter 16, provides an overview of the steps involved in designing a wind project and estimating its long-term average energy production.

In most chapters, discussion questions aimed at classroom use and recommendations for further reading are provided.

1.5 QUESTIONS FOR DISCUSSION

1. What is the principal cause of pressure gradients (differences) between different points on the earth's surface? When a pressure gradient increases, how does the wind tend to change?

2. What are the three principal mechanisms affecting the speed and direction of the wind near the earth's surface?

3. Give two examples of mesoscale atmospheric circulations. Does either of these mechanisms occur in the country or region where you live? If so, where?

4. What would have a larger surface roughness length, a grass field or a forest? All other things being equal, how is this difference likely to affect wind speeds at the height of a wind turbine?

5. What is turbulence, and on what timescales does it occur? How does turbulence affect wind turbine output?
6. What is wind shear, and why is it important?
7. What parameters of the wind resource need to be known to estimate the energy production of a single turbine? Why is it important to know the predominant wind direction when designing a wind power plant with more than one turbine?
8. How is air density related to the amount of energy that can be generated from the wind?

SUGGESTIONS FOR FURTHER READING

Ahrens CD. Essentials of meteorology. 5th edn. USA: Brooks Cole; 2007. p. 504.

Brock FV, Richardson SJ. Meteorological measurement systems. USA: Oxford University Press; 2001. p. 304.

Burton T, Sharpe D, Jenkins N, Bossanyi E. Wind energy handbook. New York: John Wiley & Sons, Inc.; 2001. p. 642.

Garratt JR. The atmospheric boundary layer. New York: Cambridge University Press; 1992. p. 336.

Lutgens FK, Tarbuck EJ, Tasa D. The atmosphere: an introduction to meteorology. 11th edn. USA: Prentice Hall; 2009. p. 508.

Robinson P, Henderson-Sellers A. Contemporary climatology. 2nd edn. USA: Prentice Hall; 1999. p. 352.

Stull RB. Meteorology for scientists and engineers. 2nd edn. USA: Brooks Cole; 1999. p. 528.

Wallace JM, Hobbs PV. Atmospheric science: an introductory survey. 2nd edn. USA: Academic Press; 2006. p. 504.

PART 1

WIND MONITORING

2

OVERVIEW OF A WIND RESOURCE ASSESSMENT CAMPAIGN

A wind resource assessment campaign, like other technical projects, requires careful planning and coordination guided by a clear set of objectives. It is often constrained by tight budgets and schedules. Its ultimate success rests on the quality of the program's assembled assets, namely, sound siting and measurement techniques, trained staff, high quality equipment, and appropriate data analysis and modeling techniques.

This chapter provides an overview of the design and implementation of a wind resource assessment campaign and identifies where these concepts are discussed in the rest of the book. It should be noted that wind resource assessment is just the first step in the life cycle of a utility-scale wind energy project. Other major steps not covered in this book, shown in Figure 2-1, include permitting, financing, construction, operation, and decommissioning. To learn more about wind energy projects beyond the resource assessment phase, the reader is encouraged to consult resources such as those listed in the "Suggestions for Further Reading" section.

A wind resource assessment campaign can be divided into three main stages: site identification, resource monitoring, and resource analysis.

Wind Resource Assessment: A Practical Guide to Developing a Wind Project, First Edition.
Michael Brower et al.
© 2012 John Wiley & Sons, Inc. Published 2012 by John Wiley & Sons, Inc.

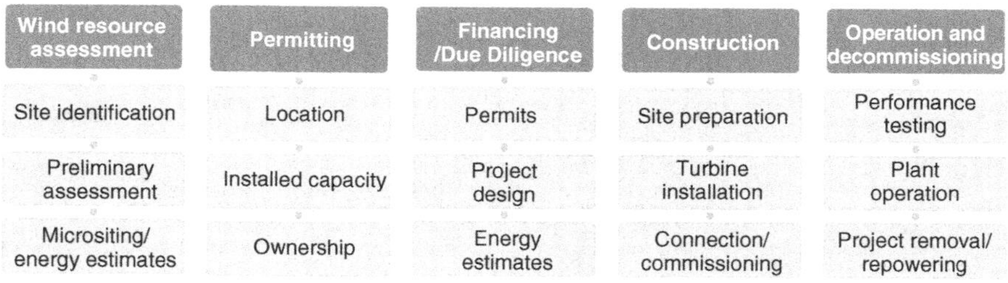

Wind resource assessment	Permitting	Financing /Due Diligence	Construction	Operation and decommissioning
Site identification	Location	Permits	Site preparation	Performance testing
Preliminary assessment	Installed capacity	Project design	Turbine installation	Plant operation
Micrositing/ energy estimates	Ownership	Energy estimates	Connection/ commissioning	Project removal/ repowering

Figure 2-1. The life cycle of a utility-scale wind energy project. *Source:* AWS Truepower.

2.1   SITE IDENTIFICATION

The first stage of the wind resource assessment campaign identifies one or more candidate wind energy project sites. This may involve surveying a relatively large region (e.g., a county, province or state, or country). A leading consideration is usually the wind resource, which may be estimated using wind maps and publicly available wind data. Other considerations may include market conditions, transmission access and capacity, site constructability and access, community and government support, and environmental and cultural sensitivities. Site-screening techniques and criteria are described in Chapter 3.

As a first step, it is recommended that geographic data be collected and compiled in a Geographic Information System (GIS). Once a GIS project has been created, appropriate criteria can be applied to select candidate sites in an efficient and systematic way. Another advantage of creating a GIS is that once a candidate site is selected, much of the wind monitoring campaign design and, subsequently, the wind project design can be carried out in a virtual environment.

Whether a GIS is used or not, the final site selection should be informed by site visits to confirm the physical conditions on which the selection was based (such as the condition of roads and locations of transmission lines) and to assess firsthand the political, regulatory, cultural, and other factors that may help or hinder development.

2.2   RESOURCE MONITORING

Once a candidate site is identified, the second stage involves the measurement and characterization of the wind resource. It is at this stage that wind monitoring towers are likely to be installed. The most common objectives of the monitoring are as follows:

- To verify whether a sufficient wind resource exists to justify further investigation.
- To compare and rank the wind resources between different candidate sites.
- To obtain representative data for estimating the performance and economic viability of different wind turbine models.
- To provide a sound basis for wind resource analysis.

Chapters 3–8 provide information and recommendations to support this phase of the wind resource assessment process. They include guidelines for designing and carrying out a complete measurement program.

2.2.1 Wind Monitoring Campaign Design

The general objective of a wind monitoring campaign is to obtain the best possible understanding of the wind resource from the top to the bottom of the turbine rotor and across the project area, consistent with the project's budget and schedule. This is achieved by placing meteorological towers and ground-based remote sensing systems in appropriate locations and obtaining a sufficient amount of data to characterize the resource. Chapters 3–5 provide guidance in designing a wind monitoring campaign using tall towers; Chapter 8 discusses remote sensing systems.

Tower Number and Placement. The main goal when deciding how many towers to install and where they should be placed within the project area is to minimize the uncertainty in the wind resource at potential turbine locations. Meeting this objective calls not just for monitoring where the wind is strongest but capturing the full diversity of resources, from the best to the worst, likely to be experienced by the turbines. The size of the area, topography, land cover, and other factors come into play in making this decision. Recommendations regarding the number and placement of meteorological masts are provided in Chapter 3.

Instrument Height. Measuring the wind resource at the turbine hub height (and preferably through the entire rotor plane), rather than extrapolating measurements from lower heights, reduces the uncertainty in energy production estimates. The choice of height depends on a number of factors including project size, tower cost, local regulations (e.g., aviation-related height restrictions), and knowledge of the site's wind shear. If the shear is well understood, the value of very tall towers is reduced; on the other hand, where the shear is difficult to characterize, such towers may be very cost-effective. For large wind projects (>100 MW), it is recommended that at least one in three meteorological towers be of at least hub height. Additional information on hub-height and taller towers can be found in Chapter 3.

Tower Instrumentation. The main task of the monitoring program is the collection of accurate wind speed, wind direction, and air temperature data. Wind speed data are the most important indicator of a site's wind resource. Multiple measurement heights are needed to determine a site's wind shear. Wind direction frequency information is important for optimizing the layout of wind turbines within a wind farm and for carrying out wind flow and wake modeling. Air temperature measurements provide additional information about the site conditions and help determine air density.

Recommendations for standard instrumentation packages are discussed in detail in Chapters 4 and 5. These sections also outline optional additions to the typical instrument package, which can be implemented if consistent with the project's goals and budget. These options demonstrate the need for a detailed campaign design that takes all project variables into account.

Ground-Based Remote Sensing. Sodar and lidar, two relatively recent additions to the technologies available for measuring wind speed, can be useful for spot-checking the wind resource at different points within the project area and for measuring the wind profile throughout the rotor plane. Short-term (4–12 weeks) campaigns are typical, but longer or multiple campaigns may be advisable for large projects (>100 MW), in complex terrain, or for projects where significant seasonal variation of shear is expected. More information on remote sensing technology can be found in Chapter 8.

2.2.2 Measurement Plan

Common to all monitoring programs is the need for a measurement plan. Its purpose is to ensure that all facets of the wind monitoring program combine to provide the data needed to meet the program's objectives. It should be documented in writing and reviewed and accepted by the project participants before being implemented. The plan should specify the following elements:

- measurement parameters (e.g., speed, direction, temperature);
- equipment type, quality, and cost;
- equipment monitoring heights and boom orientations;
- number and locations of monitoring masts;
- minimum desired measurement accuracy, duration, and data recovery;
- data sampling and recording intervals;
- parties responsible for equipment installation, maintenance, data validation, and reporting;
- data transmission, screening, and processing procedures;
- quality control (QC) measures;
- data reporting intervals and format.

It is generally recommended that wind monitoring last at least 1 year (12 consecutive months), although a longer period produces more reliable results, and that subsequent masts that are installed overlap the first in time. The data recovery for all measured parameters should be as high as possible, with a target for most tower sensors of at least 90%, with few or no extended data gaps. The rate actually achieved will depend on a number of factors, including the remoteness of the site, weather conditions, the type and redundancy of instruments, and methods of data collection.

2.2.3 Monitoring Strategy

At the core of the monitoring strategy are good management, qualified staff, and adequate resources. It is best if everyone involved understands the roles and responsibilities of each participant and the lines of authority and accountability. Everyone should be familiar with the program's overall objectives, measurement plan, and schedule. Communication among the participants should be frequent and open.

It is recommended that the project team include at least one person with field measurement experience. Data analysis, interpretation, and computer skills are also important assets. Available staff and material resources must be commensurate with the measurement program's objectives. High standards of data accuracy and completeness require appropriate levels of staffing, an investment in high quality equipment and tools, prompt response to unscheduled events (e.g., equipment outages), access to spare parts, routine site visits, and timely review of the data.

Two components that are integral to the monitoring strategy are station operation and maintenance and the data collection process.

Station Operation and Maintenance. Ongoing maintenance and careful documentation of each wind resource monitoring station is necessary to preserve the integrity and achieve the goals of the measurement campaign. It is recommended that a simple but thorough operation and maintenance plan be instituted. This plan should incorporate various quality assurance measures and provide procedural guidelines for all program personnel. Specific recommendations for the operation and maintenance of wind resource monitoring stations are provided in Chapter 6.

Data Collection and Handling. The objective of the data collection and handling process is to ensure that the data are available for analysis and protected from corruption or loss. Chapter 7 provides background information about how data are stored locally at the monitoring station and how they can be retrieved and protected. Suitable data transmission documentation is also described.

2.2.4 Quality Assurance Plan

An essential part of every measurement program is the quality assurance plan, an organized and detailed action agenda for guaranteeing the successful collection of high quality data. The quality assurance plan should be prepared in writing once the measurement plan is completed.

1. *Quality Assurance Policy*. The program manager should establish and endorse the quality assurance plan, thus giving it credibility for all personnel.
2. *Quality Assurance Coordinator*. The link between the plan and the program management is the quality assurance coordinator. This person should be familiar with the routine requirements for collecting valid data. If the quality assurance plan is to be taken seriously, he or she must be authorized to ensure that all personnel are properly trained, correct procedures are followed, and corrective measures are taken in the event of problems. In addition, the coordinator should maintain the proper documentation in an organized format.

Data quality is usually measured by representativeness, accuracy, and completeness. The quality assurance plan relies heavily on the documentation of the procedures involved to support claims of data quality. It is recommended that the quality assurance plan include the following components:

- equipment procurement tied to the program's specifications;
- equipment calibration method, frequency, and reporting;
- monitoring station installation, verification, and operation and maintenance checklists;
- data collection, screening, and archiving;
- data analysis guidelines (including calculations);
- data validation methods, flagging criteria, reporting frequency, and format;
- internal audits to document the performance of those responsible for site installation and operation and maintenance and for data collection and handling.

Another goal of quality assurance is to minimize the uncertainties that unavoidably enter at every step of the measurement processes. No tower perfectly represents the entire area it represents, no sensor measures with perfect accuracy, and no data gathered over a limited period perfectly reflect conditions a wind plant may experience during its lifetime. However, if the magnitude of these uncertainties is understood and controlled through a concerted quality assurance plan, the conclusions can be properly qualified to provide useful information.

2.3 WIND RESOURCE ANALYSIS

The third stage of the wind resource assessment campaign entails the description of the wind resource at all relevant temporal and spatial scales to support the optimal placement of turbines within the project area and the most accurate possible estimation of energy production. Part 2 of this book, encompassing Chapters 9–16, deals with this stage, including data validation, characterization of the observed resource, adjustments for wind shear and long-term wind climate, numerical wind flow modeling, project design and energy production calculations, and uncertainty.

2.3.1 Data Validation

Once the data from the monitoring system have been successfully transferred to an office computing environment, the data must be checked for errors and validated. The completeness and reasonableness of the data are assessed, and invalid or suspect values are flagged. This process also serves to detect potential problems with the instrumentation or data logger. Recommended data validation procedures are described in Chapter 9.

2.3.2 Characterizing the Observed Wind Resource

After the wind resource data have been validated, they are analyzed to generate a variety of statistics that help characterize the site's wind resource. Common statistics include mean speed, speed and direction frequency distributions, shear, turbulence intensity, and wind power density. A description of these metrics and associated equations is provided in Chapter 10.

2.3.3   Estimating the Hub Height Resource

Since meteorological towers are often shorter than a turbine's hub height (the center of the rotor), where its power curve is defined, it is often necessary to extrapolate speed measurements between heights. This task requires a careful and often subjective analysis of information about the mast and site, including the observed shear, local meteorology, topography, and land cover. The information in Chapter 11 helps guide the analyst through this process.

2.3.4   Climate Adjustment

The objective of climate adjustment is to correct measurements taken over a limited period to long-term historical conditions. This is important because wind speeds can vary substantially from the norm even over a period of a year or longer. A process known as *measure, correlate, predict* (MCP) is typically used to relate and adjust on-site measurements to a long-term reference. This reduces the uncertainty in energy production estimates. Chapter 12 provides an overview of the MCP process.

2.3.5   Wind Flow Modeling

Since on-site wind measurements are usually limited to just a few locations within a project area, wind flow modeling, most often done with computer software, must ordinarily be used to estimate the wind resource at all locations where wind turbines might be deployed. Chapter 13 provides an overview of the types of wind flow models that are available, their inputs and appropriate applications, and the uncertainties and challenges associated with them.

2.3.6   Uncertainty in Wind Resource Assessment

A sound understanding of the uncertainty associated with the wind resource assessment process is necessary for project financing. Chapter 15 reviews the potential sources of uncertainty and how they are estimated and provides typical ranges of uncertainty values.

2.3.7   Project Design and Energy Production

The last stage is to design the project and estimate its energy production. This often complicated process is usually performed with specialized software, which, starting from the results of the numerical wind flow modeling, allows the user to rapidly test different turbine layouts and to arrive at one that maximizes energy production. The software also calculates losses resulting from the wakes cast by turbines. Chapter 16 provides an overview of the project design and energy production estimation process.

SUGGESTIONS FOR FURTHER READING

Gipe P. Wind Energy Basics Revised: A Guide to Home- and Community-scale Wind Energy Systems. USA: Chelsea Green Publishing; 2009.

Windustry. Wind Energy Basics series. Available at http://www.windustry.org/wind-basics/learn-about-wind-energy/learn-about-wind-energy.

Wizelius T. Developing wind power projects: theory and practice. UK: Earthscan; 2007. Available at http://books.google.com/books?id=eTaNk1VaQTYC.

SITING A WIND PROJECT

Siting a wind project generally involves a number of steps, including selecting one or more candidate wind project sites within a large region, identifying locations within each site for the installation of wind monitoring systems, and permitting and leasing the land.

Since the region to be surveyed is sometimes quite large (as large as a state, province, or country), the site-selection process should be designed to efficiently focus on the most suitable areas. This chapter discusses siting criteria and some widely used tools and techniques. It also outlines steps to be taken following the initial site identification, including field surveys, choosing appropriate tower locations, obtaining permits for tower installation, and entering option agreements with landowners.

3.1 SITE SELECTION

The following are among the leading factors that might be considered in selecting a wind project site.

Wind Resource Assessment: A Practical Guide to Developing a Wind Project, First Edition.
Michael Brower et al.
© 2012 John Wiley & Sons, Inc. Published 2012 by John Wiley & Sons, Inc.

1. *Wind Resource.* This is almost always a key consideration because the better the resource, the greater the potential power production and project revenues. Before a wind monitoring campaign is conducted, the developer must consider sources of regional wind resource information to identify potentially attractive sites. This process is described in Section 3.2.

2. *Buildable Windy Area.* The larger the area where turbines can be installed with an adequate resource, the larger the wind project can be. The buildable windy area is often constrained by topography and also by other factors described below.

3. *Proximity to Existing Transmission Lines.* The costs and risks associated with building new lines to connect wind projects to the existing transmission grid are substantial. In general, wind project developers try to minimize the distance that must be covered.

4. *Road Access.* The developer should consider whether it is feasible to transport wind monitoring equipment to the site by truck through the existing roads and trails and the possible need to build new roads or upgrade existing roads to support the eventual delivery of wind turbines.

5. *Land Cover.* All other things being equal, development costs are generally greater in forested terrain than elsewhere, as trees have to be cleared for wind monitoring masts and eventually for wind turbines, service roads, and other plant needs. Conversely, some land cover types, such as cropland and rangeland, may be especially conducive to wind energy development.

6. *Land Use Restrictions.* Areas may be off-limits for a variety of reasons, such as for military use or wildlife protection. Such restrictions may eliminate a site from consideration or constrain its buildable area.

7. *Proximity to Residential Areas.* A common concern for communities is the proximity of wind turbines to residential areas. Residents may fear the turbines will generate too much noise or create a blight on the landscape. In many regions, these concerns are codified in the form of required setbacks from existing homes and other buildings.

8. *Cultural, Environmental, and Other Concerns.* These issues can extend well beyond officially designated restricted areas. For example, some sites may be especially important to particular groups for historical or religious reasons. Others may encroach on sensitive wildlife habitats not under official protection. Yet others may be deemed by the local community to have exceptional scenic or esthetic value. Although issues like these may not strictly rule out development, they can arouse public opposition to a project and thereby impede the development process and increase costs. The developer should be aware of objections that could be raised and take them into account in the site selection.

With the appropriate data, most of the factors described above are amenable to quantitative analysis using a GIS. Such systems have become an integral part of today's site-selection process, as they enable analysts to efficiently organize and analyze a large amount of information and to screen sites against a number of often competing criteria. For example, given the necessary inputs, a GIS could be queried to find all sites outside defined exclusion zones and within a specified distance of existing transmission lines, encompassing a contiguous area sufficiently large to support a wind project of a certain size, and with a specified minimum mean wind speed. After some experimentation with the criteria, the analyst can usually hone in on a "short list" of candidate sites meeting the developer's requirements.

The most useful geographic data to incorporate into a GIS during this phase typically include the following:

- wind resource maps
- topographic data (digital elevation or terrain model)
- land cover data (classified by vegetation type or use)
- water bodies
- administrative boundaries
- excluded areas (natural parks, military zones, urban areas, etc.)
- buildings and other structures requiring setbacks
- roads, railroads, and paths
- transmission lines and substations
- pipelines (natural gas, oil)
- radar and airspace restrictions
- competing or neighboring projects.

Appendix B lists some useful sources of global GIS data. Additional sources of data for particular regions and countries may be available as well. In some cases, the information can be digitized (converted to GIS format) from aerial or satellite photographs.

A GIS cannot address every potential factor or concern, nor are the data inputs always accurate. For these reasons, it is prudent to follow up a GIS-based site screening with field visits, as described in Section 3.3.

3.2 REGIONAL WIND RESOURCE INFORMATION

Obtaining information about the wind resource in a region is a key step in the site-selection process. Although usually insufficient by itself to determine a project's feasibility, such information can suggest the range of performance to be expected by wind projects that might be built in the region and can point toward potential sites. Two common sources of regional wind resource information are wind resource maps and publicly available wind measurements.

3.2.1 Wind Resource Maps

Regional wind resource maps can be a useful starting point for identifying attractive wind project sites. Aside from allowing users to survey a large region at a glance, most offer the benefit of being compatible with GIS. Since the 1990s, there has been an explosion in the availability of regional wind resource maps. Some regions, such as North America and parts of Europe, have been mapped multiple times by different companies and groups using a variety of methods (Fig. 3-1). Some sources of wind resource maps are listed in Appendix B.

However, regional resource maps must be used with caution, as their accuracy and spatial resolution vary widely. Some maps provide little more than general guidance about the wind resource in a region (and may be based on questionable methods and data), while others may offer sufficient accuracy and detail to be used for preliminary site selection and plant design. It is important for the analyst to investigate the methods and data used in creating a particular map and to determine whether it has been compared against independent, high quality wind measurements and, if so, the error margins obtained.

The types of information presented in wind resource maps also vary. Some maps indicate the estimated long-term mean wind speed, while others indicate the expected mean wind power density in watts per square meter of swept rotor area. Neither parameter can be translated directly into production by a wind turbine, which depends also on other factors such as the speed frequency distribution and air density, as well as on the specific turbine model and hub height. Some wind map vendors provide such supplemental information on request, including estimates of capacity factor for particular turbine models.

Even the best regional wind resource maps are usually not accurate enough to replace on-site measurements, although exceptions can sometimes be made for small wind projects where the added precision of on-site measurements does not justify the cost of a wind monitoring campaign. Quoted uncertainties in the mean speed typically fall in the range of several tenths to one meter per second or more, and the spatial resolution ranges from 100 m to 5 km. Confidence in the maps is usually greatest in relatively simple terrain and where ample validation data provided by high quality measurements exist. A greater uncertainty can be expected in complex terrain and in data-sparse regions.

3.2.2   Wind Measurements

Publicly available wind data can be useful for assessing the wind resource in a region, especially if the wind monitoring stations are in locations that are representative of sites of interest for wind projects. An example would be a tower on a ridge line that runs parallel to a similar ridge under consideration. Tall towers instrumented specifically for wind energy assessment are greatly preferred, but airports and other weather stations can provide a helpful indication of the wind resource as well. It is important in all cases to obtain as much information as possible about each station to determine whether or not the data are reliable. Several elements should be considered in this determination:

- station location
- tower type and dimensions

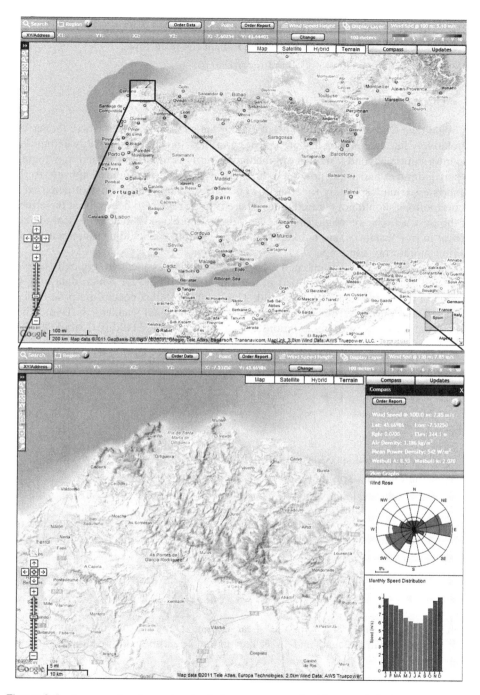

<u>Figure 3-1.</u> An example of a regional wind resource map, in this case of Spain. Some online maps provide additional information such as wind roses and patterns of seasonal variation, as shown in the lower map panel. *Source*: AWS Truepower.

- surrounding topography, obstacles, and surface roughness
- sensor heights, boom orientations, and distances from tower
- sensor maintenance protocol and records
- duration of data record
- QC and analysis applied to the data.

Wind data tend to be more representative of the surrounding area where the terrain is relatively flat. In complex terrain or near coastlines, the ability to reliably extrapolate the information beyond a station's immediate vicinity is more limited and may require expert judgment and wind flow modeling. Even in flat terrain, good exposure to the wind is essential, especially for short towers. Measurements taken in obstructed areas or on rooftops should not be used unless there is good reason to believe that the effects of the obstructions are small. In any event, data from existing meteorological towers, except in rare instances, are unlikely to be able to replace on-site measurements from a wind monitoring campaign.

These days, towers instrumented for wind resource assessment are typically at least 50 m tall, whereas towers installed in the 1980s and 1990s, for which data may be available, were often only 20–30 m tall. The heights of most general-purpose meteorological towers range from 3 to 20 m, and the international standard is 10 m. When comparing data from different stations, mean wind speeds should be extrapolated to a common reference height (e.g., 80 m, a typical wind turbine hub height). This can be done conveniently using the power law:

$$v_2 = v_1 \left(\frac{h_2}{h_1}\right)^{\alpha} \tag{3.1}$$

where

$v_2 =$ the projected speed at the desired height h_2
$v_1 =$ the observed speed at the measurement height h_1.
$\alpha =$ a nondimensional wind shear exponent.

The uncertainty in the projected speed depends on both the ratio of heights and the uncertainty in the wind shear exponent, as can be seen in the following equation (derived from Eq. 3.1):

$$\sigma_v = 100\sigma_\alpha \ln \left(\frac{h_2}{h_1}\right) (\%) \tag{3.2}$$

Here, σ_α is the uncertainty[1] in the shear exponent and σ_v the uncertainty in the projected speed in percent, and ln is the natural logarithm. For example, suppose

---

[1] Here and elsewhere in this book, the uncertainty is defined as the *standard error*, σ, which is the standard deviation of the distribution of possible results of a measurement. For normally distributed data, a measurement will fall within one standard error of the mean of the distribution about 68% of the time.

Table 3-1. Typical ranges of wind shear exponents for various types of terrain and land cover

Terrain type	Land cover	Approximate range of annual mean wind shear exponent
Flat or rolling	Low to moderate vegetation	0.12–0.25
Flat or rolling	Patchy woods or forest	0.25–0.40
Complex, valley (sheltered)	Varied	0.25–0.60
Complex, valley (gap or thermal flow)	Varied	0.10–0.20
Complex, ridgeline	Low to moderate vegetation	0.15–0.25
Complex, ridgeline	Forest	0.20–0.35
Offshore, temperate	Water	0.10–0.15
Offshore, tropical	Water	0.07–0.10

Source: AWS Truepower.

the uncertainty in the shear exponent at a particular location is judged to be 0.10 (e.g., $\alpha = 0.20 \pm 0.10$), then the uncertainty in the mean speed extrapolated from 30 to 80 m would be 10%. (Note that this is the uncertainty associated with the shear alone; it does not count the uncertainty associated with the speed measurement or the instrument height.)

For most publicly available data sets, the wind shear exponent is not known and, even if published, may not be accurate. Wind shear exponents vary widely depending on vegetation cover, terrain, the characteristics of the atmosphere, and other factors. Table 3-1 presents typical ranges of annual mean shear exponent in different regions and climates. It is important to note that these estimates assume the tower in question is taller than the surrounding vegetation or obstacles. Where this is not the case, it may be impossible to extrapolate the measurements with any assurance. Where there is doubt, a wind resource analyst with experience in the region should be consulted.

Ideally, data sets should span at least 1 year of measurement to reduce the effect of seasonal and interannual variations and should provide consistent data for at least 90% of that period. A useful format is a time series of hourly or 10-min wind speed and wind direction measurements, which can be analyzed for a number of wind characteristics such as diurnal and seasonal patterns. In some instances, wind data summaries may be available. Although convenient, such summaries should be used with caution unless the analyst is familiar with the QC procedures and analytical methods and is confident they were correctly applied. Otherwise, it is usually best to perform one's own analysis from the original, raw data.

3.3 FIELD SURVEYS

It is recommended that all candidate wind project sites be visited in person. Three main goals of the visits are (i) to confirm the assumptions and data used in a GIS-based site screening (such as the presence and location of existing roads and transmission lines), (ii) to obtain additional information not available in a map or GIS format, and (iii) to select places to install wind monitoring systems. The following items are typically documented during a site visit:

- accessibility to the site
- potential visual and noise concerns (e.g., notable scenic value, nearby residences)
- issues of potential cultural, environmental, historical, or other community sensitivity
- locations of significant obstructions that may affect wind observations
- possible wind monitoring locations, including site coordinates, accessibility, and surroundings (discussed in Section 3.4)
- cellular telephone service reliability for automated remote data transfers.

The site evaluator should refer to a detailed topographic map of the area to plan the trip and note pertinent features to be visited. A Global Positioning System (GPS) should be used during the visit to record the exact location (latitude, longitude, and elevation) of each point of interest. It may be especially convenient to link the GPS to a laptop running GIS software. A video or still camera record of the visit is also helpful not only for the site screening but also for the subsequent analysis and interpretation of the wind resource data once monitoring is under way. When assessing possible tower locations, the evaluator should gauge whether trees need to be cleared to provide a sufficiently large, open area for tower erection. Also, if a guyed meteorological tower is to be installed, the soil conditions should be determined so the proper anchor type can be chosen. (For more information about the installation of monitoring masts, see Chapter 5.)

Field visits also provide an opportunity for the developer to become acquainted with landowners, community representatives, business leaders, government officials, and others who may be involved in or affected by the proposed wind project or who may have a say in its approval. The monitoring program's objectives can be presented in a friendly, face-to-face conversation, and questions and concerns can be noted and addressed, if possible. Although wind turbines are accepted in many regions, environmental, cultural, visual, noise, permitting, and other issues can still pose important challenges. It is in the developer's interest to investigate and address these challenges as early as possible in the siting process, before making a large investment of time and money. For more information on these issues, consult the "Suggestions for Further Reading" section.

3.4 TOWER PLACEMENT

There are two distinct types of monitoring towers: dedicated towers installed specifically for wind resource monitoring and preexisting towers.

3.4.1 Dedicated Towers

Several guidelines should be followed when choosing the locations for new, dedicated monitoring towers:

- The towers should be placed as far away as possible from significant obstructions that would not be representative of obstructions at likely turbine locations.
- For small projects, a location should be selected that is representative of where most of the wind turbines are likely to be sited, not necessarily where the best wind is to be found.
- For large projects, a diverse set of locations representing the full range of conditions where wind turbines are likely to be sited should be chosen. If only one mast is installed to start with, it should be in the most representative location for the planned turbine array.
- The towers should be placed away from transmission lines and buried gas lines or electric cables, among other hazards.

One approach to tower placement is to keep the distance between any proposed turbine and the nearest tower within specified limits. With this method, it is necessary to envision a specific turbine layout (or at least its outlines) before siting the towers. While there is no clear industry standard, the guidelines in Table 3-2 may be useful.

Distance is not the only criterion that should be considered, however. It is equally important that the mast locations be representative of the terrain in which the turbines will eventually be installed. For example, it is not unusual for turbines to be placed not only along the crest of a ridge where the wind is usually strongest but also some distance down the slope. It would be beneficial in such a situation to place one or more

Table 3-2. Recommended maximum distances between monitoring masts and turbines based on terrain complexity and land cover

Project site	Terrain and land cover	Maximum recommended distance between any proposed turbine location and nearer mast, km[a]
Simple	Generally flat with uniform surface roughness	5–8
Moderately complex	Examples include inland site with gently rolling hills, coastal site with uniform distance from shore, single ridgeline perpendicular to prevailing wind	3–5
Very complex	Steep geometrically complex ridgelines, coastal site with varying distance from shore, or heavily forested	1–3

[a] The meteorological mast is assumed to be located within the proposed turbine array.
Source: AWS Truepower.

towers off the ridgeline as well. Careful attention to tower placement can substantially reduce the wind flow modeling uncertainty and consequently the uncertainty in the predicted plant energy production (Chapter 13).

Siting a tower near significant obstructions such as buildings, rock outcroppings, or isolated stands of trees can adversely affect the analysis of the site's wind characteristics (unless the proposed turbines would experience similar obstructions). Figure 3-2 illustrates the effects of an obstruction, which include reduced wind speed and increased turbulence. The zone of increased turbulence can extend up to 2 times the obstacle height in the upwind direction, 10–20 times the obstacle height in the downwind direction, and 2–3 times the obstacle height in the vertical direction. As a guideline, if sensors must be placed near an obstruction, they should be located at a horizontal distance of no less than 20 times the height of the obstruction in the prevailing wind direction.

When placing wind monitoring systems near or within forests or extensive tree stands, it is important to consider whether the vegetation is typical of where turbines are likely to be located. If so, then so long as the necessary clearances (e.g., for mast installation or sodar operation) are respected, there is no reason to avoid them. However, the lowest speed sensors on the tower should be placed well above the tree canopy to ensure an accurate measurement of wind shear. Tower instrumentation guidelines are provided in Chapter 4.

Although highly recommended for large projects and in areas of substantial shear uncertainty, the use of very tall towers (i.e., hub height and greater) can create challenges for tower placement. They often require a larger cleared area to erect than shorter towers. In forested areas, clearing requirements may limit either the tower height or the locations where towers can be placed; the same goes for steep terrain. Taller towers are also more exposed to severe weather than shorter towers and must be designed to withstand expected occurrences of icing, high winds, lightning, and

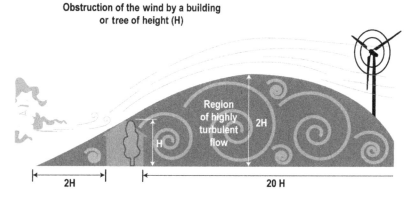

Figure 3-2. Obstruction effects on airflow. *Source*: AWS Truepower.

other potentially damaging conditions, which may increase their cost. They may also require special aviation permits or lighting.

3.4.2 Existing Multi-Use Tall Towers

At first glance, existing multi-use tall towers might appear to be a convenient and cost-effective alternative to new, dedicated towers. No new structure is required and often on-site power is already available for running existing equipment. However, such towers of opportunity may also have significant drawbacks that make accurate wind resource assessment more difficult. Points to consider when selecting an existing tall tower for a monitoring campaign include the following.

- If the tower is well outside the proposed turbine array, its value to the monitoring campaign will likely be limited.
- If the tower is unusually wide or it carries a lot of equipment such as communications dishes and radar repeaters, it may be difficult to obtain an accurate reading of the free-stream wind speed.
- The number of instruments that can be mounted and their heights and boom lengths may be constrained by the tower owner or by structural considerations.
- It may not be possible to access the tower whenever necessary to replace or repair wind monitoring equipment.
- The tower owner may install equipment at a later date that disrupts the continuity of measurements; however, it should be possible to guard against this concern in the tower lease agreement.

For these reasons, in practice, few wind projects are assessed solely or mainly on the basis of data collected from existing multi-use towers.

3.5 PERMITTING FOR WIND MONITORING

Government-issued permits may be required before a wind monitoring tower can be installed. Permitting requirements vary from country to country and even within countries, and sometimes change over time, so it is recommended that the permitting process and requirements be thoroughly researched early in the planning for the monitoring campaign.

If the mast is to be installed on public or government-owned land, it will be necessary to apply for permission from those agencies controlling access to and activities on the land. Even on privately owned land, it is not unusual for a variety of regulating agencies, from local to national, to establish permitting procedures for structures as tall as wind monitoring towers can be. Civil aviation authorities, for example, may establish a height above which a permit must be obtained; in the United States that height is 61 m (200 ft). A permit may also be required if the structure will be under an aircraft flight path or within a certain distance of an airport. Proximity to sites of particular cultural, environmental, historical, religious, or other significance may trigger other permitting hurdles.

In some countries, local authorities (e.g., town governments) exert considerable control over the use of both public and private lands within their jurisdictions, independent of national and provincial agencies. Some localities may simply require notification prior to installation of tall structures on private land. Others may require a Professional Engineering (PE) stamp on structure drawings and anchoring system designs and a detailed explanation of how the tower will be decommissioned (removed) at the end of the monitoring period. Since tilt-up towers usually fall in the category of temporary structures, it is often easier to obtain permits for them than for fixed towers.

3.6 LAND LEASE AGREEMENTS

Once a potential monitoring site is identified, the developer typically enters into an agreement with the landowner (or whatever entity controls the land) to gain access to the property for the duration of the monitoring program and to secure rights to the land should the project go forward.

This agreement often takes the form of an option. The option period typically lasts 3–5 years to allow sufficient time for the developer to install the mast and evaluate the wind resource. Before the term is over, the developer has the choice of exercising the option to lease the land for a wind project, requesting an extension, or letting the option expire. In this way, both the landowner's and developer's interests are protected during the option period. The developer is assured that the land will be available if the project goes forward, without having to purchase it or lease it for a long period in case it does not. The landowner is assured that if the project is not built, he or she will be able to offer the land to another developer or put it to another use.

During the option period, the developer usually pays a fee to the landowner for the right to place wind monitoring equipment on the site and sometimes to compensate for lost income and construction- or service-related disruptions. The compensation varies widely depending on local custom, the wind resource, the length of the option period, the desirability of the land for wind development, and the income or opportunities that may be lost from alternative uses.

Example terms include, but are not limited to, the following:

Area Leased. The lease should clearly state where the meteorological towers can be located and the total area they will occupy. Any required setbacks from residences and property lines should be stated.

Access. The developer should be able to access the monitoring equipment to retrieve data and carry out repairs and maintenance in a timely fashion, with appropriate notice given to the landowner.

Approved Uses. The lease or option agreement should specify what uses the landowner reserves for the land around the monitoring equipment. For instance, the landowner may reserve the right to continue to grow crops or raise cattle.

Crop Protection. Typical lease provisions require developers to use their best efforts to minimize damage and to compensate landowners for any damage that may occur. Mitigation measures covered in the lease agreement may include soil preservation or decompaction to remedy the effects of project-related vehicle traffic.

Liability and Insurance. The agreement should contain provisions to protect landowners from liability arising from accidents. The agreement should also require that the developer carry a general liability insurance policy.

Term and Decommissioning. The duration of the option period should be clearly stated. When the agreement ends, the developer is usually responsible for removing the tower and restoring the site to a suitable condition.

Compensation Payment Schedule. The agreement should outline how the landowner will be compensated and the payment schedule.

3.7 QUESTIONS FOR DISCUSSION

1. For the region where you live, list and give a short description of the main factors that might affect the selection of a site for a wind project. Rank the considerations in the order of decreasing importance, state whether and how they can be quantified, and identify possible sources of information to support them.

2. The wind resource is usually one of the key considerations in selecting a wind project site. Can you imagine conditions in which it might be much less important? Consider such factors as the variation in the resource across the region and the types of incentives that might be available for wind energy projects.

3. You have been hired to design a wind monitoring campaign for a 100-MW wind project in the region where you live. What trade-offs do you need to consider in determining (i) the number of towers to be installed, (ii) their height and instrumentation, and (iii) their placement within the project area?

4. Investigate and discuss permitting procedures for wind energy projects in your region. What government agencies are likely to be involved at the local, state or provincial, and national levels? Do these agencies have experience with wind projects? Consider just one of the agencies. Describe the kind of permit that would be required and the procedure for obtaining it.

5. Using the Internet, find a sample land lease or option agreement for a wind energy project. Consider and discuss its applicability for a specific type of landowner (e.g., a farmer) in a specific region.

SUGGESTIONS FOR FURTHER READING

American Wind Energy Association. Wind energy siting handbook. Prepared by Tetra Tech EC, Inc., and Nixon Peabody LLP. Feb 2008. Available at http://www.awea.org/sitinghandbook/download_center.html.

National Wind Coordinating Committee. Permitting of wind energy facilities: a handbook. Aug 2002. Available at http://www.nationalwind.org/assets/publications/permitting2002.pdf.

New York State Energy Research and Development Authority (NYSERDA). Wind energy toolkit. Prepared by AWS Truewind, LLC. May 2009. Available at http://www.nyserda.ny.gov/en/Page-Sections/Renewables/Large-Wind/Wind-Energy-Toolkit.aspx.

4

MONITORING STATION INSTRUMENTATION AND MEASUREMENTS

Measurements taken by meteorological instruments mounted on tall towers are the foundation of most wind resource assessment. Each of these instruments is designed to record a specific environmental parameter. The basic parameters essential for any wind monitoring program are the horizontal wind speed, wind direction, and air temperature. Other, optional parameters may be collected as well depending on the priorities and budget of the wind resource monitoring program. This chapter describes the common types of meteorological instruments, along with the recorded parameters, sampling intervals, and desired measurement accuracy; associated data loggers, storage devices and transfer equipment; and power supplies, tower types, and wiring.

4.1 BASIC MEASUREMENTS

4.1.1 Horizontal Wind Speed

Wind speed is the most important indicator of a site's wind resource. Obtaining accurate readings of the free-stream wind speed (i.e., the speed unaffected by the tower, instruments, and other station components) over a representative period is, therefore,

Wind Resource Assessment: A Practical Guide to Developing a Wind Project, First Edition.
Michael Brower et al.
© 2012 John Wiley & Sons, Inc. Published 2012 by John Wiley & Sons, Inc.

Table 4-1. Specifications for basic sensors

Specification	Anemometer (wind speed)	Wind vane (wind direction)	Temperature probe
Measurement range	0–50 m/s	0°–360° (≤8° deadband)	−40° to −60°C
Starting threshold	≤1.0 m/s	≤1.0 m/s	N/A
Distance constant	≤3.0 m	N/A	N/A
Operating temperature range, °C	−40 to 60	−40 to 60	−40 to 60
Operating humidity range, %	0–100	0–100	0–100
System error	≤1% (at 1σ)	5°–10°	≤1°C
Recording resolution	≤0.1 m/s	≤1°	≤0.1°C
Lifetime (service interval), yr	2	2–6	2–6

N/A = not applicable. *Source:* AWS Truepower.

Table 4-2. Basic measurement parameters

Measurement parameters	Example heights (60-m tubular tower)	Example heights (83-m lattice tower)
Wind speed, m/s	57.2, 47.4, and 32.0 m	80, 60, and 40 m
Wind direction, degrees	53.5 and 43.7 m	80, 60, and 40 m
Temperature, °C	3 m	3 m and/or hub height

Source: AWS Truepower.

the top priority of any wind monitoring campaign. Achieving this goal requires careful attention to the choice of instruments, mounting configuration, and tower design. Multiple redundant anemometers and measurement heights are strongly encouraged to maximize data recovery and to accurately determine a site's wind shear. Sensor-mounting recommendations and diagrams of typical monitoring configurations are provided in Section 5.7.

Three anemometer types are used for the measurement of horizontal wind speed. Of these, the cup anemometer is the most popular because of its low cost and generally good accuracy. However, both propeller and sonic anemometers are used in some settings.

- *Cup Anemometer.* This instrument consists of three or four cups connected to a vertical shaft. The wind causes the cup assembly to turn in a preferred direction. A transducer in the anemometer converts this rotational movement into an electrical signal, which is sent through a wire to the data logger. The logger measures the frequency or magnitude of the signal and applies a predetermined multiplier (slope) and offset (intercept) to convert the signal to a wind speed (Fig. 4-1).
- *Propeller or Prop-Vane Anemometer.* This consists of a propeller mounted on a horizontal shaft that is kept pointing into the wind by a tail vane. Like a cup anemometer, a propeller anemometer generates an electrical signal whose frequency or magnitude is proportional to the wind speed. This type of anemometer

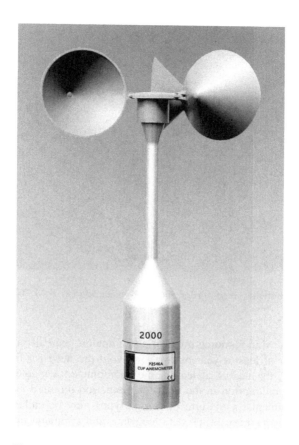

Figure 4-1. A cup anemometer. *Source:* WindSensor.

can record slightly lower speeds than cup anemometers under turbulent conditions. This so-called underspeeding is caused by the prop-vane's tendency to oscillate around the central direction or to lag behind sudden wind direction shifts, with the result that the propeller does not always point directly into the wind (Fig. 4-2).

- *Sonic Anemometer.* This instrument, which does not have any moving parts, measures the wind speed and direction by detecting variations in the speed of ultrasound (sound waves whose frequency is above the range of human hearing) transmitted between fixed points. Some sonic anemometers measure wind in two dimensions, whereas others measure in three. Because they have no rotational inertia, sonic anemometers are more responsive to rapid speed and direction fluctuations than cup or propeller anemometers. They are also usually more expensive than other types and require more power (Fig. 4-3).

When selecting an anemometer type and model, the following factors should be considered.

Figure 4-2. A propeller anemometer and wind vane. *Source:* R.M. Young Company.

- *Durability.* A wind resource monitoring campaign generally involves collecting wind data for at least a year or two. To avoid the need for frequent and costly replacements, the use of at least some anemometers capable of surviving and holding their calibration in the field for the period required is recommended. In some environments, a mixture of sensor types may be called for to achieve a balance between survivability, data recovery, and accuracy. In an extended wind monitoring program, provision should be made for the regular inspection and replacement of anemometers.

- *Operating Environment.* Not every anemometer is suited to every environment. Conditions that may cause problems include icing, heavy rain, lightning, sand and dust, extreme temperatures, and saltwater intrusion. Extreme conditions can cause anemometers and direction vanes to read incorrectly or stop working altogether. Heated anemometers are available from most manufacturers to cope with icing, and it is recommended that at least one or two be installed on every mast where significant icing is expected. This will reduce data loss. Heated anemometers are discussed further in Section 4.2.

- *Starting Threshold.* This is the minimum wind speed at which the anemometer starts and maintains rotation. Since low wind speeds are of no interest in wind energy generation, the starting threshold for most anemometers on the market is adequate for wind resource assessment. The exception is anemometers designed to measure vertical wind speeds, which must be sensitive to small departures (both positive and negative) from zero.

- *Distance Constant.* This is a measure of how long an anemometer takes to respond to an abrupt change in wind speed. It is defined as the distance that must

Figure 4-3. A sonic anemometer. *Source:* Campbell Scientific.

be traveled by a cylindrical volume of air passing through the anemometer for the anemometer to record 63% of an instantaneous speed change. Anemometers with relatively large distance constants may overestimate the mean wind speed in turbulent conditions compared to other anemometers. This is because they tend to respond more quickly to a rise than to a drop in speed. Sonic anemometers are not susceptible to this overspeeding effect. Unheated anemometers commonly used for resource assessment have distance constants ranging from 1.8 to 3.0 m, while heated anemometers, which are bulkier because of their heating elements, tend to have much larger distance constants.

- *Response to Vertical Wind.* In relatively steep terrain, the wind often has a significant vertical component. Since wind turbines are sensitive only to the horizontal component of the speed, and turbine power curves are defined and measured under horizontal flow conditions, the vertical component should ideally be ignored. Different anemometers have different characteristics in this respect. The easiest to use are sonic and propeller anemometers. Mounted correctly, sonic anemometers provide a direct measure of the horizontal wind components, whereas propeller anemometers, like wind turbines, are sensitive only to the horizontal wind. Some cup anemometers (known as *2D anemometers*) behave rather like propeller anemometers and have little or no sensitivity to the vertical component of the wind. Others (known as *3D anemometers*) respond to vertical winds and thus can produce a misleading estimate of the horizontal speed. Corrections can be made for these anemometers if the vertical wind speed can be measured or estimated and the anemometer's sensitivity to the inclination angle is known (1).

* *Sensor Calibration.* The transfer function (slope and offset) for cup and propeller anemometers either can be a default (or consensus) function previously established by testing a large number of sensors of the same model or can be measured specifically for the sensor that was purchased. In the latter case, the sensor is said to be calibrated. Either approach may be acceptable depending on the circumstances. There is evidence that for NRG #40 cup anemometers, in particular, the accepted consensus function produces results that match anemometers used for power performance testing more closely than do measured transfer functions (2, 3). An advantage of using calibrated sensors, whether the measured transfer function is used or not, is that there is greater assurance that "bad" sensors will be discovered before they are installed in the field. In addition, with calibrated sensors, it is possible to determine the change in sensor response over the course of the monitoring period by removing it from the field and testing it repeatedly. High quality, undamaged anemometers should exhibit very little change. Calibrated anemometers are sometimes assigned a lower uncertainty than uncalibrated ones.

There are many wind-industry-accepted cup anemometers on the market. A list is provided in Appendix A. Although each of the sensor models meets wind industry standards, it may be desirable to deploy more than one model on each mast. This strategy can reduce the risk of data losses or measurement errors caused by problems affecting just one model. Some anemometers have been classified according to the standards outlined by institutions such as MEASNET or the International Electrotechnical Commission (IEC).[1] The performance of these anemometers complies with the institution's specifications for high accuracy applications such as power curve testing. However, it is usually unnecessary for every anemometer on a mast to meet this standard.

4.1.2 Wind Direction

Wind direction measurements are a necessary ingredient for modeling the spatial distribution of the wind resource across a project area and for optimizing the layout

---

[1] The IEC standard on power performance measurements (IEC 61400-12-1) classifies cup anemometers based on sensor accuracy. It should be noted that this document requires that turbine performance tests be carried out with calibrated Class I anemometers and that the measured calibration constants be used.

of the wind turbines. A wind vane is usually used to measure wind direction (with prop-vane and sonic sensors, no separate vane is required, though one or two may be desirable for redundancy). In the most familiar type, a horizontal tail connected to a vertical shaft rotates to align with the wind (Fig. 4-4). To define the wind direction with adequate redundancy, it is recommended that wind vanes be installed on at least two monitoring levels. Ideally, they should not be mounted on the same booms or even at the same heights as the anemometers, as they could interfere with obtaining accurate speed readings. It is customary to mount the direction vanes 1 or 2 m below the anemometers.

For a wind vane with a potentiometer-type transducer, the data logger usually provides a voltage across the potentiometer's entire resistive element and measures the voltage where the potentiometer's wiper arm makes contact. The ratio between these two voltages determines the position of the vane with respect to its reference direction. The potentiometer cannot cover a full 360°, however, and a small gap is left between where it starts and ends. In this so-called deadband, the output signal floats at random and the direction cannot be determined. The best practice is to orient the deadband in an infrequent direction, and usually directly facing the tower; in addition, the size of the deadband should not exceed 8°. The precision of the wind vane direction is another important consideration. A resolution better than or equal to 1° is recommended.

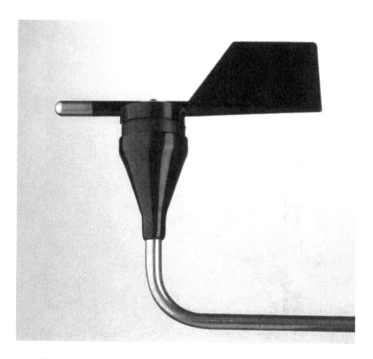

Figure 4-4. A direction vane. *Source:* NRG Systems, Inc.

4.1.3 Air Temperature

Air temperature is an important characteristic of a wind farm's operating environment. It is normally measured 2–3 m above ground level or near hub height, or at both levels. In most locations, the average air temperature near ground level is within 1°C of the average at hub height. Air temperature is used to estimate air density, which affects the calculation of power production. The air temperature readings are also utilized in the data validation process to detect icing.

An ambient air temperature sensor is typically composed of three parts: the transducer, an interface device, and a radiation shield (Fig. 4-5). The transducer contains a material (usually nickel or platinum) exhibiting a known relationship between resistance and temperature. Thermistors, resistance thermal detectors, and temperature-sensitive semiconductors are common element types. The resistance value is measured by the data logger (or other interface device), which then calculates the air temperature based on the known relationship. The temperature transducer is housed within a radiation shield to prevent it from being warmed by direct sunlight.

Figure 4-5. A temperature sensor. *Source:* Campbell Scientific.

4.2 ADDITIONAL MEASUREMENTS

Depending on the site conditions and the needs, priorities, and budget of the monitoring program, additional sensors may be included to measure vertical wind speed, high accuracy temperature (for determining vertical temperature gradients), relative humidity, barometric pressure, and solar radiation. Table 4-3 lists the nominal specifications for these sensors; the measurement parameters associated with each sensor are summarized in Table 4-4. Bear in mind that each additional instrument requires power, and there are limits to the number of instrument channels supported by data loggers.

4.2.1 Vertical Wind Speed

In complex terrain (defined by the IEC as having a slope of more than 10% within a distance of 20 times the hub height from turbines (4)), it is recommended that anemometers capable of measuring the vertical wind speed be used in conjunction

Table 4-3. Specifications for optional sensors

Specification	Pyranometer (solar radiation)	Vertical propeller anemometer	High accuracy temperature	Barometer (atmospheric pressure)
Measurement range	0–1500 W/m²	−50 to 50 m/s	−40 to 60°C	94–106 kPa (see level equivalent)
Starting threshold	N/A	≤1.0 m/s	N/A	N/A
Distance constant	N/A	≤4.0 m	N/A	N/A
Operating temperature range, °C	−40 to 60	−40 to 60	−40 to 60	−40 to 60
Operating humidity range, %	0–100	0–100	0–100	0–100
System accuracy	≤5%	≤3%	≤0.1°C	≤1 kPa
Recording resolution	≤1 W/m²	≤0.1 m/s	≤0.01°C	≤0.2 kPa

N/A = not applicable. *Source:* AWS Truepower.

Table 4-4. Optional measurement parameters

Measurement parameters	Typical monitoring heights
Vertical wind speed, m/s	2 m below top anemometer height
Delta temperature, °C	3 m above ground, and 2 m below top anemometer height
Barometric pressure, kPa	2–3 m above ground
Relative humidity, %	3 m above ground, or 2 m below top anemometer height
Solar radiation, W/m²	3–4 m above ground

Source: AWS Truepower.

with standard instruments. By directly measuring the vertical component of the wind, the energy-producing horizontal component can be better estimated. In addition, vertical wind speed measurements can be an important input for turbine loading and suitability calculations, as severe or frequent off-horizontal winds can cause damaging loads and wear.

Two common approaches for measuring the vertical wind speed are to mount a propeller anemometer with its axis pointed vertically (Fig. 4-6) or to use a sonic 3D anemometer. Whichever approach is taken, since vertical winds can vary greatly with height above ground, it is recommended that the anemometer be placed as close to hub height as possible without causing interference with other sensors (a vertical separation of 1−2 m between sensors is usually adequate).

Since vertical motions are often very small, an anemometer of unusual sensitivity is called for. A propeller anemometer requires a transducer that can indicate both upward and downward motions. The signal is usually a direct current (DC) voltage whose sign and magnitude are interpreted by the data logger (or an interface device). Sonic sensors are more expensive than propeller anemometers but offer the advantage of measuring both vertical and horizontal wind components simultaneously and at precisely the same height and position.

Figure 4-6. A vertical propeller anemometer. *Source:* R.M. Young Company.

4.2.2 Heated Anemometers

The build up of ice can cause anemometers to rotate more slowly, stop, or even break because of the load (e.g., falling ice). Likewise, it can cause wind vanes to be off-balance or alter their aerodynamic profile, distorting the directional readings, or can freeze them in one position. For these reasons, where frequent or heavy icing is expected, met towers should be equipped with at least one or two heated anemometers and direction vanes. Unfortunately, heated sensors consume much more power than unheated sensors (power supply options are discussed in Section 4.7).

Except during periods of icing, heated anemometers are generally less accurate than unheated ones. Therefore, it is recommended that unheated anemometers be the primary source of wind data, while heated ones should be used only to fill gaps in the primary data record. In addition, to maintain as much consistency as possible in the heated-anemometer readings, it is recommended that power be applied to the heating elements throughout the year, not just in the cold season. In a typical configuration, an unheated anemometer might be paired with a heated anemometer on each of the top two levels of the mast, and one of the two direction vanes on the mast might be heated. This approach strikes a balance between high overall data recovery and good measurement accuracy.

4.2.3 Delta Temperature

The parameter ΔT (pronounced delta tee) is the difference in temperature between two heights above ground. It is a measure of the thermal stability, or buoyancy, of the atmosphere. The challenge is to measure the temperature difference with sufficient accuracy to be useful. According to the EPA Quality Assurance Handbook (1989), the maximum allowable ΔT error is 0.003 C/m (degrees Celsius per meter of height). With heights of, say, 10 and 40 m, the allowable error is just 0.1°C. To achieve this, a pair of identical temperature-sensing subsystems calibrated and matched by the manufacturer is typically used. To maximize the height difference and thus the precision of the result, one sensor should be placed about 3 m above the ground and the other a short distance (e.g., 2 m) below the top sensing level on the tower. In addition, both sensors need to be mounted and shielded in the same manner so they respond similarly to ambient conditions. To further reduce errors, a radiation shield that uses either forced (mechanical) or natural (passive) aspiration is required (to meet EPA guidelines, forced aspiration may be necessary). The data logger manufacturer should be consulted to determine compatible sensor types and models.

4.2.4 Barometric Pressure

Knowing the barometric pressure along with the air temperature can help improve the accuracy of air density estimates, as normal variations in pressure at the same temperature can affect air density by about 1%. However, barometric pressure is difficult to measure accurately in windy environments because of the dynamic pressures induced by the wind passing across the instrument enclosure. For this reason, high accuracy instruments are quite expensive, and as a result, most resource assessment programs do not measure barometric pressure, instead they rely either on temperature and elevation alone or on pressure readings from a regional weather station. Under most conditions, both methods can yield acceptable accuracy. The main exception is projects at especially high elevations (such as >2000 m above sea level) and with no nearby weather station at a similar altitude, where it is recommended that high accuracy air pressure measurements be taken.

Several barometric pressure sensors, or barometers, are commercially available (Fig. 4-7). Most models use a piezoelectric transducer that sends a DC voltage to a data logger and may require an external power source. Consult with the data logger manufacturer to determine a compatible sensor model. Note that the transducer needs to be exposed to the ambient outside air pressure. It must not be mounted in an airtight enclosure or in a way that wind flow around the inlet could induce pressure changes.

4.2.5   Relative Humidity

Since the amount of water vapor in the air affects its density, the use of a relative humidity sensor can improve the accuracy of air density estimates. However, the humidity effect is usually small, so this parameter is rarely measured for this purpose. In cold climates, a relative humidity sensor is sometimes used for icing analysis.

Figure 4-7. A barometric pressure sensor. *Source:* Campbell Scientific.

4.2.6 Global Solar Radiation

The solar energy resource can be measured as part of a wind monitoring program. Solar radiation, when used in conjunction with wind speed and time of day, can also be an indicator of atmospheric stability.

A pyranometer is used to measure global horizontal (total) solar radiation, which is the combination of direct sunlight and diffuse sky radiation striking a horizontal plane. One common type of pyranometer uses a photodiode, which generates a very small current proportional to the amount of incident sunlight (called *insolation*). Another type uses a thermopile, a group of thermal sensors, which produces a very small voltage. The data logger (or a supplementary interface device) applies a predetermined multiplier and offset to calculate the global solar radiation reading. Since the output signal from the sensor is usually very small (microamps or microvolts), it may have to be amplified to be read by the logger.

The pyranometer must be leveled to measure global horizontal solar radiation accurately. When installed on a tower in the northern hemisphere, it is best to locate the sensor on a boom extending southward, above or beyond any obstructions to minimize shading from other instruments and the tower; the reverse applies, of course, in the southern hemisphere. The recommended measurement height is 3–4 m above ground. Pyranometers may require frequent maintenance visits for cleaning and releveling.

4.3 RECORDED PARAMETERS AND SAMPLING INTERVALS

It is highly recommended that the parameters to be measured and their sampling and recording intervals conform with typical wind industry practice. Adherence to these standards will facilitate the wind resource analysis and any subsequent external review. The industry standard recording interval is 10 min, although occasionally other (usually shorter) intervals are used. The parameters are generally sampled once every 1 or 2 s (depending on the logger model) within each 10-min interval (to avoid errors, the sampling frequency should not be greater than the pulse frequency from the anemometer). Depending on the parameter, the data logger records interval averages, standard deviations, and maximum and minimum values. Data recording should be serial in time, with all records marked by a time and date stamp. These requirements are all standard functions of data loggers designed for wind energy applications. It should be noted that the time stamp for some data loggers refers to the preceding interval, while for others it refers to the following interval.

The recorded values are the basis for the data validation procedures described in Chapter 9. Each is presented below and summarized in Table 4-5.

4.3.1 Average

The average or mean value in each 10-min interval is recorded for all parameters except wind direction. For the direction, the average is defined as a *vector resultant value*, which is the direction implied by the means of the northerly and easterly

Table 4-5. Derived statistics for basic and additional parameters

Measurement parameters	Recorded values
Wind speed, m/s	Average, standard deviation, min/max
Wind direction, degrees	Average, standard deviation, max gust direction
Temperature,°C	Average, min/max
Solar radiation, W/m^2	Average, min/max
Vertical wind speed, m/s	Average, standard deviation, min/max
Barometric pressure, kPa	Average, min/max
Delta temperature,°C	Average, min/max
Relative humidity, %	Average, min/max

speeds. Averages are used in reporting wind speed variability, as well as wind speed and direction frequency distributions.

4.3.2 Standard Deviation

The standard deviation should be determined for both wind speed and wind direction and is defined as the *population standard deviation* (σ) for all 1- or 2-s samples within each 10-min interval. The standard deviations of wind speed and wind direction are indicators of turbulence. They are also useful for detecting suspect or erroneous data.

4.3.3 Maximum and Minimum

The maximum and minimum values observed during each interval should be recorded for all parameters. This is especially important for the maximum 3-s gust (speed), which can affect whether a particular turbine model is deemed suitable for the site. If possible, the coincident directions corresponding to the maximum and minimum wind speeds should also be recorded.

4.4 DATA LOGGERS

Data loggers (or data recorders) have evolved from dials and strip charts read by a person to a variety of digital stand-alone devices. Many manufacturers now offer complete data-logging systems that include integrated data storage and transfer options (Fig. 4-8).

All data loggers store data locally, and many can transfer the data to another location through cellular telephone, radio frequency (RF) telemetry, or satellite link. Remote data transfer allows the user to obtain and inspect data without making frequent site visits and also to verify that the logger is operating correctly. Section 4.6 provides detailed information on data transfer equipment options.

The data logger must be compatible with the sensor types employed and must be able to support the desired number of sensors, measurement parameters, and sampling

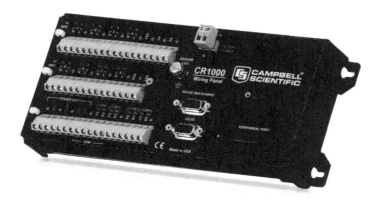

Figure 4-8. A data logger. *Source:* Campbell Scientific.

and recording intervals. It is prudent to mount the logger in a noncorrosive, water-tight, lockable enclosure to protect the logger and the peripheral equipment from the environment, theft, and vandalism. It is recommended that the data logger

- be capable of storing data values in a sequential format with corresponding time and date stamps;
- contribute negligible errors to the signals received from the sensors;
- have an internal data storage capacity of at least 40 days;
- possess an onboard real-time clock so that the time stamps will remain accurate even if the logger loses power;
- operate in the same environmental extremes as those listed in Table 4-1;
- offer retrievable data storage media when a remote uplink is not possible;
- offer remote data collection options;
- operate on battery power (which may be augmented by other sources such as a solar panel);
- offer nonvolatile memory storage so that data are not lost if power fails.

A number of electronic data loggers that meet these criteria are commercially available, and a vendor list is provided in Appendix A.

4.5 DATA STORAGE DEVICES

Every data logger contains a computer running on operating system software. It includes a small data buffer to temporarily hold data for processing. The computer accesses this buffer to calculate the desired parameters, such as means and standard deviations. The resulting data values are then stored in memory. Some data loggers have a fixed, or firm, operating system that cannot be altered or can be only slightly modified; others are user-interactive and can be reprogrammed for different tasks. In older models, the operating system and data buffers are sometimes stored in volatile memory. Their drawback is that they need a continuous power source to retain data. Data loggers that incorporate internal backup batteries or use nonvolatile memory are preferred because data are less likely to be lost.

4.5.1 Data Processing and Storage

Data processing and storage methods vary according to the data logger (Chapter 7). There are two commonly used formats for recording and storing data: ring memory and fill and stop.

1. *Ring Memory.* Once the available memory is filled to capacity, the newest data record is written over the oldest.
2. *Fill-and-Stop Memory.* Once the memory is filled to capacity, no additional data are archived. This stops the data logging.

In the past, the ring memory format was preferred over fill-and-stop memory because it allowed data logging to continue if the operator was unable to retrieve the data before the memory buffer was filled. Given the memory storage capacity of modern data loggers, however, this is of much less concern. Today's memory buffers are typically able to store at least 6–12 months of data, unless the recording interval is much less than the usual 10 min.

4.5.2 Storage Devices

Most manufacturers offer several options for data storage devices. The most common are presented in Table 4-6.

4.6 DATA TRANSFER EQUIPMENT

The selection of a data transfer and handling process and data logger model depends on the monitoring program's resources and requirements. As a rule, the manufacturer should be consulted to ensure compatibility between system components. It is also recommended that a unit be purchased in advance for testing before committing to a new monitoring system configuration.

Data are typically retrieved and transferred to a computer either manually or remotely.

Table 4-6. Data storage devices

Storage device	Description	Download method/needs
Memory card	Independent memory chips in numerous formats (e.g., MMC, SD, microSD, SDHC, Memory Stick, USB flash drive) used in cameras and other devices	Read and erased on-site or replaced. Reading device and software required
Solid-state module	Integrated electronic device that directly interfaces with the data logger	Read and erased on-site or replaced. Reading device and software required
Data card	Programmable read–write device that plugs into a special data logger socket	Read and erased on-site or replaced. Reading device and software required
EEPROM data chip	An integrated circuit chip incorporating an electrically erasable and programmable read-only memory device	EEPROM reading device and software required
Magnetic media	Familiar floppy disk or magnetic tape (i.e., cassette)	Software required to read data from the media
Portable computer	Laptop or notebook type computer	Special cabling, interface device, and/or software may be required

4.6.1 Manual Data Transfer

Manual retrieval requires visiting the site to transfer data. Typically this involves two steps:

1. The current storage device (e.g., data card) is removed and replaced and sent to another location for download. Alternatively, the data can be transferred at the site to a laptop computer. Many loggers use an RS-232 serial port to interface with a computer. Computers that do not have an RS-232 port can use a USB port and USB/RS-232 adapter.

2. The collected data are transferred to a central computer where the data are analyzed and backed up.

The main disadvantage of manual data transfer is the need for frequent site visits to ensure that all the equipment is operating correctly. If a sensor or logger malfunctions between visits, data that would have been collected from that point forward until the next visit may be lost. In addition, there is an increased risk of data mishandling (e.g., lost or damaged storage cards). For these reasons, most wind monitoring campaigns these days rely on remote data transfer.

4.6.2   Remote Data Transfer

Remote transfer requires a telecommunications link between the data logger and the central computer. The communications system may employ direct-wire cabling, telephone lines, cellular telephone, RF telemetry, or satellite-based telemetry, or for redundancy, a combination of these components. The main advantage of this method is that the data can be retrieved and inspected more frequently (e.g., weekly) than might be practical with site visits. This means that problems with the equipment can be more quickly identified and resolved, thus likely increasing data recovery. Many logger manufacturers now offer integral remote data collection equipment. The main disadvantage of the remote method is the cost of the equipment. In addition, some sites have poor cellular coverage, and other, noncellular options can be expensive.

There are two methods of remote data retrieval: those initiated by the recipient (call out) and those initiated by the logger (phone home). The first type requires the recipient to oversee the telecommunication operation. Steps include initiating the call to the in-field logger, downloading the data, verifying data transfer, and erasing the logger memory. Some call-out data logger models are compatible with computer-based terminal emulation software packages with batch calling. Batch calling automates the data transfer process, allowing the user to download data from a number of monitoring sites at prescribed intervals. Batch programs can also be written to include data verification routines. The data logger manufacturer should be consulted to determine the compatibility of its equipment with this feature.

A phone-home data logger automatically calls the home computer at prescribed times to transfer data. In the past, the phone-home method could not be used to support as many towers as the call-out method because call times had to be spaced far apart to allow for slow or repeated transfer attempts. The newest generation of data loggers solves this problem by using the Internet to send data out as attached e-mail files. This allows for concurrent data transfer from multiple sites. In addition, the data can be delivered to more than one computer, providing greater data security and convenience.

Data loggers with remote data transfer via cellular communications are gaining popularity because of their ease of use and reasonable cost. The cellular signal strength and type (GSM or CDMA) at the site should be determined in advance; this can be done with a portable phone. Where the signal strength is weak, an antenna with higher gain can sometimes be successful. Failing that, a satellite modem linking to the Globalstar or Iridium network is an option.

Guidelines for establishing a cellular account are usually provided by the data logger supplier. It is important to work closely with the equipment supplier and cellular telephone companies to resolve any questions before monitoring begins. It is best to schedule data transfers during off-peak hours to take advantage of discounted rates and faster network speeds.

4.7   POWER SOURCES

All electronic data loggers require a power source, which should be sized to meet the total power requirements of the monitoring system. The leading power options are described below.

4.7.1   Household Batteries

The newest generation of loggers employ low power electronic components whose operation can be sustained by common household batteries (D cells, 9 V, and others) for 6 months to a year. Although the systems are generally reliable, if the batteries fail, data collection stops. In addition, the power is not sufficient for towers with heated sensors or for other special power needs. To address these issues, logger batteries are often augmented by another power source.

4.7.2   Solar Battery Systems

For more reliable long-term operation and for meeting larger power needs, the most popular choice is a rechargeable lead-acid battery coupled to a solar panel. Packaged solar battery systems are offered by most logger vendors for this purpose.

Lead-acid batteries are a good choice because they can withstand repeated discharge and recharge cycles without greatly affecting their energy storage capacity, and they can hold charge well in cold temperatures. Caution should always be used when working with large batteries like these to avoid a short circuit between the battery terminals. It is also recommended that newer battery designs which encapsulate the acid into a gel or paste to prevent spills, called *nonspillable* or *gel batteries*, be used. Although long-lived, lead-acid batteries subjected to many charge and discharge cycles eventually lose capacity. It is important to take this decline into account when sizing the battery if it is intended to last in the field for several years.

The solar panel must be large enough to operate the monitoring system and keep the battery charged during the worst expected conditions (usually in winter). To avoid outages that may cause data loss, it is recommended that the solar and storage system be designed for at least 7 days of autonomous operation (without recharging). The solar system must also be reverse-bias protected with a diode to prevent power drain from the battery at night. In addition, it must include a voltage regulator to supply a voltage compatible with the battery and to prevent overcharging during months with the most sunlight.

Logger vendors offering solar battery packages can advise on the proper size for any location.

4.7.3   AC Power

Alternating current (AC) power is not normally required for wind monitoring systems. Moreover, it is unusual (except for communications towers) for a met mast to be close enough to a source of AC power to make connecting to it worthwhile. Nevertheless, where AC power is conveniently at hand, the instrumentation loads are unusually large, or solar panels are not practical, then AC power can be the right choice. It should be used only to trickle-charge a storage battery, not to power the logger directly. A surge/spike suppression device should be installed to protect the system from electrical transients. In addition, all systems must be properly tied to a common earth ground (Chapter 5).

4.7.4   Other Power Options

Other power sources that may be used in some circumstances include small wind systems, wind/solar hybrid systems, and diesel generators. Small wind and wind/solar hybrid systems can be a good choice where there is plenty of wind and limited solar radiation (e.g., in arctic environments) or where solar panels are likely to be blocked by trees or other obstacles most of the time. Diesel generators require refueling, but there are sites where they are practical.

4.8   TOWERS AND SENSOR SUPPORT HARDWARE

4.8.1   Towers

Towers for wind resource monitoring come in several varieties. There are two basic structural types: tubular and lattice. Tubular towers consist of a single hollow pole or several segmented hollow poles connected end to end, whereas lattice towers consist of either three or four segmented corner poles attached by interconnecting struts. For both types, three versions are available: tilt-up, telescoping, and fixed. Tilt-up towers are mounted on a hinged base, and the unhinged end is raised into place using a winch and cable. Telescoping towers are jacked up from a vertical position. Fixed towers are generally lifted into place. In addition, towers can be either self-supporting or stabilized by guy wires anchored in the ground. Self-supporting towers require sufficient structural strength to resist lateral wind loads on their own. Guyed towers can be less massive since they derive their lateral stability from the wires.

For most sites, tubular, tilt-up towers are preferred for wind resource assessment because they are relatively easy to raise and lower (i.e., the tower can be assembled and sensors mounted and serviced at ground level), they require minimal ground preparation, and they are relatively inexpensive. However, tilt-up tubular towers are currently limited in height to 80 m, and so fixed towers (usually lattice type) must be used if a greater height is desired. In some regions, lattice towers are widely used for all heights because they can be cheaply manufactured locally.

Towers should be strong enough to withstand the extremes of wind and ice loading expected at the location and should be stable enough to resist wind-induced vibration. Tower manufacturers should be able to provide guidance on allowable environmental conditions. Note that some jurisdictions have their own design requirements for wind and ice loading, which can have implications for the permitting process. In coastal environments, resistance to saltwater exposure is important. Guy wires should be secured with anchors that match the site's soil conditions (the vendor and installer should advise on this), and all ground-level components should be clearly marked to prevent accidents. Protection from lightning (with instruments including lightning rod, cable, and grounding rod) is a must. In some locations, security measures to prevent vandalism, theft, and unauthorized climbing may be required, as well as protection against cattle or other large animals.

Further information about tower installation is provided in Chapter 5, and the special requirements of offshore towers are discussed in Chapter 14.

4.8.2 Sensor Support Hardware

Anemometers and direction vanes are most often mounted on booms that extend horizontally from the side of the tower. Near the end of the boom is attached a short vertical mast, or pillar, to hold the sensor above the boom. The boom and pillar should be long enough that the influence of both the tower and the boom on the speed and direction readings is kept small, except when the sensors are directly downwind of the tower. Chapter 5 provides guidelines for achieving this goal.

Anemometers can also be mounted vertically off the top of the tower. This has the advantage of eliminating the problem of tower wake (or shadow), although care must still be taken to use a long-enough vertical boom to minimize the influence of the tower top on the observed wind speed. Since vertically mounted anemometers experience different tower effects from horizontally mounted anemometers, the vertically mounted configuration is not suitable for measuring wind shear. One application of this configuration is turbine power performance testing, for which a pair of hub-height vertically mounted anemometers is commonly deployed.

A further consideration is that the boom and vertical mounting pillar must be fixed in such a way that the anemometer shaft is precisely vertical (for cup and sonic anemometers and direction vanes) or precisely horizontal (for prop-vane anemometers). A tilted or skewed boom or mast can result in an incorrect measurement of the free-stream wind speed or direction.

It is advisable to use sensor support hardware that is able to withstand the same wind and ice loading extremes as the tower and is not prone to wind-induced vibration. The hardware should be protected against corrosion, especially in coastal environments. Drainage holes in sensor housings need to be kept open to prevent water accumulation and expansion in freezing conditions. Tubular (hollow) sensor booms and pillars should be used instead of solid stock.

Guidance regarding instrument placement, including the heights of sensors and orientation of booms with respect to prevailing wind directions, is provided in Chapter 5.

4.9 WIRING

The following are the guidelines for selecting the proper electrical wire or cable type:

- Use the proper class wire for the voltage and sensor type

- Use wire with an ultraviolet (UV)-resistant insulating jacket
- Use insulation and conductor types that are flexible over the full temperature range expected at the site
- Use shielded and/or twisted-pair cables. Both prevent ambient electrical noise from affecting measurements. The normal practice is to tie only one end of a shielded cable's drain wire to earth ground.

Most sensor and data logger manufacturers offer sensor cables that meet these requirements and are compatible with their equipment.

4.10 MEASUREMENT SYSTEM ACCURACY AND RELIABILITY

Manufacturers use various definitions and methods to express their product's accuracy and reliability. This section provides the basic information needed to select the proper equipment and meet the specifications cited in Table 4-1.

4.10.1 Accuracy

The accuracy of any system tends to be dominated by that of its least accurate component. For most types of wind resource measurement, that component is usually the sensor. Errors associated with the electronic subsystem (data logger, signal conditioner, and associated wiring and connectors) are typically negligible.

The system error of an instrument is defined as the standard deviation of errors observed for a large number of instruments of that type, with respect to an accepted standard. For a given instrument, the measured value should be within the quoted system error of the true value with 68% confidence and within twice the system error of the true value with 95% confidence. However, the system error applies only to controlled conditions; factors that may arise in the field, such as turbulence or tower influences on the free-stream wind speed, are not considered and may be significant.

Not all manufacturers express system error in a consistent format. The error is typically expressed in one of three ways:

1. As an absolute difference (e.g., $\leq 1°C$)

$$|\text{Measured value} - \text{Accepted standard value}|$$

2. As a percentage difference (e.g., $\leq 3\%$)

$$100 \left(\frac{\text{Measured value} - \text{Accepted standard value}}{\text{Accepted standard value}} \right)$$

3. As a minimum ratio to the accepted standard value (e.g., 95% accuracy)

$$100 \left(\frac{\text{Minimum Measured value}}{\text{Accepted standard value}} \right)$$

Accuracy is often confused with precision. System precision (sometimes also expressed in terms of standard deviation) refers to the consistency of repeated values recorded by the same instrument under the same conditions. Precision may also refer to the number of digits reported by the data logger. To avoid rounding errors in subsequent analysis, it is recommended that values be recorded to one significant digit greater than the nominal precision.

4.10.2 Reliability

System reliability is the measure of a system's ability to constantly provide valid data. Vendors usually test the reliability of their equipment to determine the product's life cycle. They will often cite a mean time between failures under certain conditions. Although this information is helpful, one of the best indications of a product's reliability is the experience of other users. The vendor can be asked for references, and users can be contacted at conferences and workshops. The user also has an important role to play in ensuring system reliability by strictly following the manufacturer's guidelines for installation and operation, using redundant sensors to maximize the recovery of critical data, and implementing a comprehensive quality assurance plan.

4.11 QUESTIONS FOR DISCUSSION

1. What are the basic environmental parameters measured in all wind resource assessment campaigns? What type of devices are used to make the measurements?

2. Using the Internet, identify an anemometer that has sufficient technical information available to determine if it would be suitable for a wind resource assessment campaign. What is the accuracy of the device, and how is it expressed? What is the distance constant for the device? Would you use the device for a wind resource assessment campaign? Why or why not?

3. Using the Internet, identify a complete tower and instrumentation package for wind resource assessment, including pricing and delivery information.

4. Suppose you are measuring the wind resource at a high elevation site in very steep terrain where winters are very cold and severe. Design an instrumentation package that you think would provide the information needed for an accurate resource assessment.

5. What might be the purpose of having two anemometers at the same height on a mast? Why not do the same with other sensors, such as direction, temperature, or pressure sensors?

6. You are conducting a wind monitoring campaign in a remote region. It is 80 km (50 miles) to the nearest town over roads that are frequently impassable because of severe weather. The cellular signal at the site is weak and intermittent. What are your options for data collection, and how would you evaluate them? Would your preferred option be different if there were five towers on the site rather than just one?

7. You are conducting a wind resource assessment requiring the use of heated anemometers. Invent a scenario where the use of a gas or diesel electric generator could be the best solution to maintain the charge of the batteries powering your equipment.

REFERENCES

1. Papadopoulos KH, et al. Effects of turbulence and flow inclination on the performance of cup anemometers in the field. Boundary-Layer Meteorol 2001;101(1):77–107.
2. Hale E. Memorandum: NRG #40 transfer function validation and recommendation. Albany, New York, USA: AWS Truewind; 2010.
3. Young M, Babij N. Field measurements comparing the Riso P2546A anemometer to the NRG #40 anemometer. Seattle, Washington, USA: Global Energy Concepts; 2007.
4. International Electrotechnical Commission (IEC) IEC 61400-12-1 WInd turbine generator systems - Part 12: Wind turbine power performance testing. (First Edition 2005–12). (IEC publications can be purchased or downloaded from http://webstore.iec.ch/.)

SUGGESTIONS FOR FURTHER READING

Brock FV, Richardson SJ. Meteorological measurement systems. New York: Oxford University Press; 2001. p. 304.

Coquilla RV. Review of anemometer calibration standards. USA: OTECH Engineering, Inc.; 2009. p. 9. Available at http://otechwind.com/wp-content/uploads/CANWEA-2009-Paper.pdf. (Accessed 2012).

International Energy Agency Programme for Research and Development on Wind Energy Conversion Systems. Expert Group Study on Recommended Practices for Wind Turbine Testing and Evaluation: Topic 11. Wind Speed Measurement and Use of Cup Anemometry, Second Print. 2003. p. 60. Available at http://www.ieawind.org/task_11/recommended_pract/11_Anemometry.pdf. (Accessed 2012).

Papadopoulos KH, Stefanatos N, Paulsen US, Morfiadakis E. Effects of turbulence and flow inclination on the performance of cup anemometers in the field. Boundary-Layer Meteorol 2001;101:77–107. Available at https://springerlink3.metapress.com/content/m047808m32314343/resource-secured/?target=fulltext.pdf&sid=nlik5u45zxbzxqb2rtiiore1&sh=www.springerlink.com. (Accessed 2012).

Strangeways I. Measuring the natural environment. 2nd ed. UK: Cambridge University Press; 2003. p. 548.

5

INSTALLATION OF MONITORING STATIONS

The installation phase of the monitoring program occurs once the site selection and wind monitoring system design have been completed, all required permits have been obtained, and the necessary equipment has been acquired. This chapter provides guidelines on key installation steps, including equipment procurement, inspection and preparation, tower installation, sensor and equipment installation, site commissioning, and documentation.

5.1 EQUIPMENT PROCUREMENT

The first step in the process is to procure the equipment that will be needed to meet the objectives of the wind monitoring program, as defined in the measurement plan. This process often involves trade-offs between cost, convenience, and performance. At this early stage of project development, budgets can be tight, leading to a desire to economize in equipment procurement. But while cost is an important consideration at all times, a monitoring program that is designed with cost as a paramount concern could be unsuccessful. For example, deploying fewer than the recommended number of masts will save money, but also increase the uncertainty in the project's energy

Wind Resource Assessment: A Practical Guide to Developing a Wind Project, First Edition.
Michael Brower et al.
© 2012 John Wiley & Sons, Inc. Published 2012 by John Wiley & Sons, Inc.

production estimate. This may cost the developer far more than the initial savings when the project is eventually financed.

The procurement process typically begins with the definition of the number of towers, the tower types and heights, the desired measurement parameters, and the desired data sampling and recording intervals. The program manager can then move on to define the sensor types and quantities (including spares); the required mounting booms, cables, and hardware for each sensor; the data logger processing requirements and the number and types of data channels required (which may affect the choice of logger model and manufacturer); and the data retrieval method (manual or remote). Along the way, normal and extreme weather conditions for the sites should be investigated to ensure that the equipment will perform reliably throughout the year.

Finally, price quotes for equipment packages meeting the program's objectives should be obtained, along with information on warranties, product support, and delivery dates, and compared between suppliers. A manufacturer that provides comprehensive product support can be an invaluable resource when installing and troubleshooting the operation of a monitoring system; some even offer training courses. A list of equipment vendors is provided in Appendix A.

5.2 EQUIPMENT ACCEPTANCE TESTING AND FIELD PREPARATION

5.2.1 Acceptance Testing

Once the equipment arrives, it is advisable to check it immediately for broken or missing parts, and all system components should be thoroughly inspected and tested. The inspection findings should be documented, and components that do not meet specifications should be returned immediately to the manufacturer for replacement.

The following acceptance testing procedures are recommended:

1. *Data Logger*
   - Ground the logger before connecting sensors to prevent damage from electrostatic discharge.
   - Turn on the data logger and check the various system voltages.
   - If applicable, set up and activate the telecommunications account (cellular or satellite based) and e-mail services following the manufacturer's instructions.
   - First connect the drain wire, then connect all sensors to data logger terminals with the shielded cabling to be used.
   - Verify that all sensor inputs are operational.
   - Verify the logger's data collection and data-transfer processes.

Here is a simple test scenario: following the manufacturer's instructions, connect a sensor to the data logger and collect a sample of data at 1-min (or higher) frequency averaging interval. Transfer the recorded data from the storage device (e.g., data card) to a computer using the logger's data management software. View the data and ensure that (i) the data logger is functioning, (ii) the data transfer was successful, (iii) the storage device is functioning, and (iv) the reported values are reasonable. Repeat the above steps using remote transfer, if required.

2. *Anemometers and Wind Vanes*
   - If calibrated anemometers were purchased, consult each calibration certificate to ensure that the reported values are within normal bounds.
   - Inspect each anemometer and vane to ensure that it spins freely through a full rotation. Check for unusual friction, and listen for binding or dragging components.
   - Using the shielded cabling, and following the manufacturer's instructions, connect each sensor to the correct data logger terminal. Verify the reasonableness of each sensor's output as displayed by the data logger. For the anemometers, verify both a zero and a nonzero value by holding and then spinning the cups or propeller. For the wind vanes, verify the values at the four cardinal points: north, south, east, and west.

3. *Temperature Sensor*
   - Perform a single-point calibration check at room temperature; once stabilized, compare the sensor temperature readings to a known calibrated thermometer if available. Deviations between sensors should not exceed $1°C$.

4. *Solar Panel Power Supply*
   - Place in direct sunlight and confirm the output voltage. Note that polarity is important when connecting to the terminals of some loggers.

5. *Mounting Hardware*
   - Inspect the sensor mounting booms to ensure they are rugged and durable.
   - Inspect any welds or joints for cracks.
   - Preassemble one mount for each type of sensor to confirm that all parts are available.

5.2.2 Field Preparation Procedures

Thorough preparation before going into the field to install equipment can save time and reduces the risk of problems requiring a costly return visit.

- Assign an identification number to each monitoring site and clearly mark equipment destined for each site.
- Enter all pertinent site and sensor information in a Site Information Log (Section 5.9).
- Install the data logger's data management software on a personal computer and enter the required information.

- If desired, program the data logger in advance with the appropriate site and sensor information (slopes and offsets). Enter the correct date and time in the data logger.
- Insert the data logger's data storage card or other storage device.
- To save valuable field installation time, assemble as many components in-house as possible. For example, sensors can be prewired and mounted on their booms.
- Some sensors are fragile, so properly package all equipment for safe transport to the field.
- Pack all tools that will be needed in the field.
- Include at least one spare of each component, when practical. The number of spares depends on the amount of wear the equipment is expected to experience, as well as the expected lead time to obtain a replacement. The cost of the spare equipment should be weighed against the time and effort needed to find a replacement should the need arise.

5.3 INSTALLATION TEAM

The quality of the data collected in a wind monitoring program depends on the quality of the installation. The installation team should have experienced personnel, one of whom is clearly assigned a supervisory role. This will promote efficiency and safety. The team should also have an appropriate number of personnel for the type of tower and equipment to be installed. The installation of a 50- or 60-m tilt-up tubular tower typically requires a crew of four, including the supervisor. Labor requirements for installing lattice towers vary, and should be determined by a qualified engineer.

The personnel responsible for the site's selection may not always be involved in the installation. If this is the case, it is important that the installation team leader obtain all pertinent site information, including the latitude and longitude (verifiable with a GPS receiver), prevailing wind direction, road maps, codes or keys to any gates, and topographic maps and site photographs that precisely show the planned tower location.

5.4 SAFETY

Installing a tower is inherently dangerous. Towers and equipment can fall on people, climbers can fall from towers, and if AC power is involved or there are nearby power lines, there is a risk of electrocution. In some remote areas, even wildlife may pose a hazard. It is essential that the team leader strictly enforce safety protocols. In addition, having experienced staff, following manufacturers' recommendations, and taking common-sense precautions will reduce risks. The team should

- be trained in and abide by all applicable safety procedures;
- remain in communication with each other and with the home office;

- follow all safety guidelines provided by the tower and equipment manufacturer;
- use common sense during the installation process. For example, if there is lightning activity, postpone work until the danger has passed;
- have the proper safety equipment, including hard hats, protective gloves, eye protection, proper shoes or boots, vests for greater visibility, a first aid kit, and, if tower climbing is required, certified climbing harnesses and lanyards;
- maintain adequate hydration, use sunscreen, and wear appropriate cold-weather clothes where necessary;
- be trained in first aid and CPR (cardiopulmonary resuscitation);
- exercise caution when driving off-road to avoid accidents;
- make sure the base of the tower is at least 1.5 tower heights away from overhead power lines;
- be aware of any equipment on the tower that may be electrically live, and if possible, turn off AC power at the tower base before working on the tower;
- before digging or installing earth anchors or rods, contact the local underground facilities protection organization to identify and mark any existing hazards (e.g., buried electric or gas lines);
- inspect any existing tower, anchors, and guy cables before conducting new work;
- tension the guy wires according to the tower manufacturer's specifications;
- have at least two tower climbers, both trained in tower rescues, for lattice tower installations;
- notify local airfields when new towers are erected to ensure that the pilots are aware of the new structure, and ensure the towers are marked in accordance with local guidelines.

5.5   DETERMINATION OF TRUE NORTH

Knowing the direction of true north is essential for interpreting direction data and is also useful during the tower layout and installation. In a surprising number of monitoring programs, direction vanes and anemometers are not oriented in the correct, documented direction. This can cause significant errors in wind flow and wake modeling and result in a poor turbine layout.

Often, directional errors arise because of confusion between magnetic and true north. Magnetic north is the direction in which the north end of a compass needle points; true north is the direction along the local line of longitude to the north pole. Sometimes, the correction from magnetic north to true north is applied wrongly, and sometimes it is applied twice, once in the field and once by the data analyst. When tower installers use a magnetic compass, the risk of error can be reduced by instructing them to orient the sensors with respect to magnetic north and by correcting the readings to true north when the data are analyzed. Fortunately, these days, most GPS receivers can indicate true north, thus eliminating the need to consider magnetic north at all.

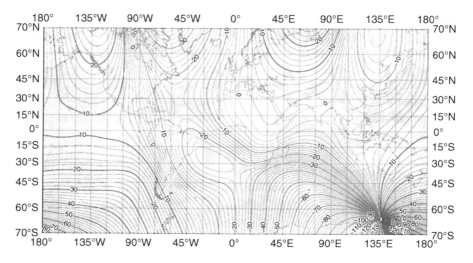

<u>Figure 5-1.</u> Map illustrating magnetic declination for the world in 2004. Blue lines indicate that magnetic north is west of true north, while red lines indicate the opposite. *Source:* US National Oceanic and Atmospheric Administration.

If a correction from magnetic north is required, the local magnetic declination (in degrees) must be established. This correction can be found on topographic or isogonic maps of the area (an example is provided in Figure 5-1). How the correction is applied depends on whether magnetic north is to the west or east of true north. To the east, the declination is expressed in degrees west of true north, and the bearing toward true north therefore equals the declination. To the west, the true north bearing is 360° minus the declination. In both cases, the true north bearing must be added to the direction relative to magnetic north to obtain the direction relative to true north. For example, suppose a sensor boom has an orientation of 150° from magnetic north. If the local magnetic declination is 15°W, then the boom orientation from true north is 135°. If, however, the magnetic declination is 15°E, then the boom orientation from true north is 165°.

5.6 TOWER INSTALLATION

5.6.1 New Tilt-Up Towers

Tilt-up towers can be erected almost anywhere, but the task is much easier if the terrain is relatively flat and free of trees. If the tower is erected on sloping or uneven ground, the guy wires may need to be adjusted often as the tower is raised. If the tower is erected in a wooded or otherwise obstructed area, there must be a clearing around the tower, which should be large enough for the tower to lie flat and the guy wires to be anchored. The required clearing for guyed tilt-up towers can be large. For example, a 60-m tilt-up tower is guyed in four directions from the tower's base. The outermost guy anchor at each corner may be 50 m (164 ft) from the base. Thus, in this example, the four anchor points form a square of roughly 71 m (233 ft) on a side. When the tower is lying flat, it extends about 10 m (33 ft)—plus the length of any lightning mast or vertical sensor boom—beyond one of the outermost anchors. This creates a kite-shaped footprint, with two sides of 71 m and two sides of at least 80 m (Fig. 5-2).

It is recommended that the guy anchors be located at four of the eight primary directions with respect to true north (i.e., N, S, E, W or NE, SE, SW, NW), as indicated by bearing reference stakes, and that one of these directions be aligned as closely as possible with the prevailing wind direction. The advantages of this strategy are, first, it is easy to verify the orientations of the sensor booms by taking a bearing from the prone mast, and second, raising the tower into (or lowering it away from) the prevailing wind direction offers a welcome degree of stability by maintaining the lifting guy wires in constant tension. Raising or lowering the tower during periods of high winds or gusts should not be attempted under any circumstances.

The tower is normally raised using a gin pole and winch-and-pulley system (Fig. 5-3). The choice of anchor for both the lifting/lowering station (where the gin pole is attached to the ground) and guy wires is critical. Anchors are most often driven into the soil (Fig. 5-4), but whether this is feasible and the type of anchor that should

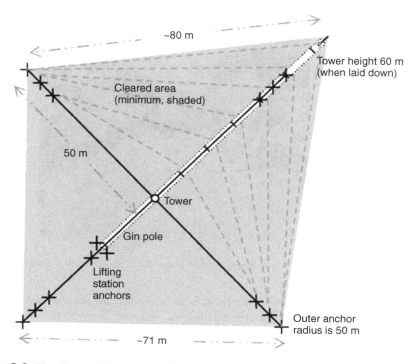

Figure 5-2. The diagram illustrates the footprint of a tilt-up tower. In this example, the prevailing wind direction is assumed to be from the southwest. The "X" marks indicate anchor points. The orange dashes represent the guy wires as the tower is being raised, and the black lines indicate the path of the guy wires when the tower is fully erected. *Source:* AWS Truepower.

Figure 5-3. Tilt-up meteorological tower being raised with a gin pole. *Source:* AWS Truepower.

Figure 5-4. Example of a screw-in guy anchor installation. *Source:* AWS Truepower.

be used depend on the subsurface characteristics at the site, which should have been determined during the initial site investigation. A mismatch between the anchor type and soil conditions could cause the anchor to fail and the tower to collapse. Note that the load-carrying capacity of the soil can vary with the weather. For example, soil that is saturated with water after a winter thaw may have much less load-carrying capacity than the same soil at other times of year.

If anchors cannot be driven into the soil (e.g., because of underground facilities or hazards), then concrete blocks can be used as counterweights. The two main disadvantages of this method are the cost of the anchors and the need to transport them to the site.

The installation of each guy anchor and lifting/lowering station anchor should adhere to the manufacturer's instructions. The lifting/lowering station anchor, which is normally connected to a winch and pulley system, warrants special attention because it must carry the entire tower load. The greatest load occurs when the tower is suspended just above the ground. This moment provides a good opportunity to evaluate the performance of the chosen anchors. If the anchors do not seem to be adequate for the soil conditions, an alternative anchoring method should be identified and tested before the tower is lifted into place. All personnel should be clear of the potential fall zone when the tower is raised.

Under proper tension, the guy wires keep the tower vertical and minimize sway. Inadequate or uneven tension can cause towers to bend or fall. The manufacturer's recommendations for guy wire tension should be followed. The installation team leader should ensure that all guy wire tension adjustments are made smoothly and in a coordinated fashion. It is also advisable to clearly mark the lower guy wires with reflective, high visibility material (such as brightly colored plastic guy sleeves) to alert pedestrians and vehicle operators. This marking should conform with state and local regulations. If large animals graze or live near the site, a fence may be necessary to protect the guy stations and tower.

5.6.2   New Lattice Towers

New lattice towers are often employed when a very tall tower is required or when there is a preference for manufacturing the towers locally. There are two basic types of lattice towers: guyed and self-supporting. Both versions are usually made of fixed-length sections connected end to end. The sections may be assembled with the tower lying flat on the ground and the tower picked up as a unit and set in place with a crane, or they may be stacked in place using a winch and jib pole system. The tower sections may all look the same, but it is important that they be installed in the correct sequence (the sections are often numbered for this purpose).

On a guyed tower, cables are attached at several heights and in at least three directions to stabilize the structure. A self-supporting tower is wider near the base to support the structure above it. Both types of tower require a solid base, usually on a concrete foundation. The guyed tower requires anchor stations located at approximately 80% of the tower's height from its base. The self-supporting type usually has three legs with a solid footing, such as a concrete pier, under each; typically, each side of its footprint is about 10% of the tower's height.

5.6.3   Existing Towers

Existing towers such as communications towers can offer several challenges. They come in a range of sizes and lattice designs, with the result that the sensor mounting hardware must often be custom designed and fabricated. The design needs must be determined during the initial site investigation; this is not a day-of-installation task.

In addition, it is recommended that each design adhere to the sensor mounting and exposure specifications presented in Section 5.7. For example, to minimize the effect of especially wide lattice towers on speed measurements, much longer mounting booms fabricated from heavier stock may be required. Where possible, anemometry should be mounted at heights where there is minimal wind flow disturbance caused by the equipment already mounted on the tower (e.g., dishes, antennas, and lightning masts).

Unlike tilt-up towers, fixed towers must be climbed for the equipment to be installed, repaired, or replaced. Before climbing is permitted, qualified personnel should evaluate the structural integrity of the tower, especially the climbing pegs, ladder, climbing safety cable, and guy wires (if present). Tower climbers must be properly trained and equipped. Since the work will be performed aloft, the weather must be given close attention. Strong wind can make it difficult to raise mounting hardware. In cold, windy weather, the danger of frostbite may be high, and tasks involving manual dexterity can become very difficult.

Note that adding support booms for anemometers and other instruments can create wind or ice loads exceeding the tower's design specifications. The implications of adding equipment to an existing tower should be reviewed by a qualified engineer.

5.7 SENSOR AND EQUIPMENT INSTALLATION

Sensors should be mounted on towers in a way that provides the desired resource information while minimizing the influence of the tower, mounting hardware, other equipment, and other sensors on the measurements. This can be achieved by adhering to the following guidelines, consulting manufacturers' instructions, and referring to the example installation configurations shown in Figures 5-5 and 5-6.

5.7.1 Anemometers

The number of heights at which the speed is measured depends partly on the height of the tower. For a 50- or 60-m tower, anemometers are typically installed at three heights, and taller towers may have four. The following general guidelines govern the selection of heights and are applicable to most towers.

- One of the heights should be as close as possible to the expected turbine hub height, consistent with other requirements.
- The topmost anemometers, if mounted on horizontal booms rather than above the tower, should be at least 10 tower diameters below the top of the tower to avoid effects of flow over the top (known as *3D flow*).
- The lowest height should be near or below the bottom of the turbine rotor plane and well above the direct influence of trees, buildings, and other obstacles; 30 m is typical.
- The heights should be as widely separated as possible to minimize uncertainty in shear, consistent with other requirements. A height ratio of at least 1.6 between

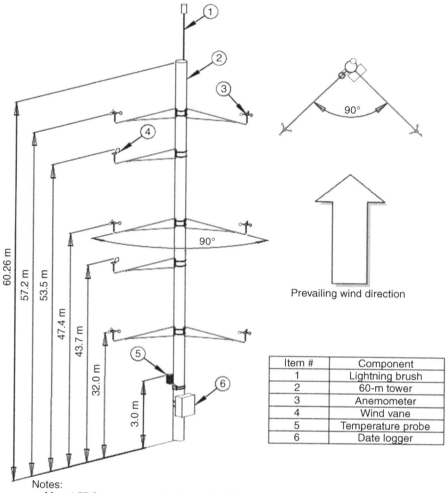

Item #	Component
1	Lightning brush
2	60-m tower
3	Anemometer
4	Wind vane
5	Temperature probe
6	Date logger

Notes:
 • Mount 57.2-m anemometer booms just above top guy ring.
 • Mount 47.4-m anemometer booms just above the second highest guy ring.
 • Mount 32-m anemometer booms 1 m above guy ring attower neck down.
 • Distances taken from the ground to the sensors (not booms).

Figure 5-5. A typical recommended mounting configuration for a 60-m tubular NRG TallTower. *Source:* AWS Truepower.

the top and bottom anemometers is recommended. For example, if the top mon-itoring height is 50 m, the lowest monitoring height could be 30 m, since the ratio of the two is 1.66.

• At projects with multiple monitoring stations (met towers or remote sensing), it is useful to match one or more of the monitoring heights between the stations to facilitate comparisons between them.

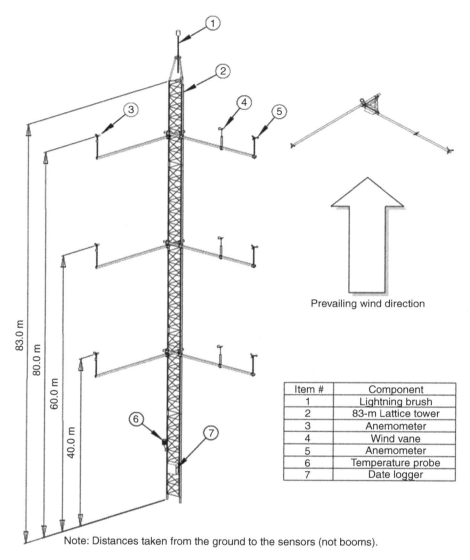

Prevailing wind direction

Item #	Component
1	Lightning brush
2	83-m Lattice tower
3	Anemometer
4	Wind vane
5	Anemometer
6	Temperature probe
7	Date logger

Note: Distances taken from the ground to the sensors (not booms).

Figure 5-6. A typical recommended mounting configuration for an 83-m guyed lattice meteorological tower. *Source:* AWS Truepower.

The following two sections show how these guidelines may be implemented for tubular and lattice towers.

Tubular Towers. The instrument heights on tubular towers are flexible but usually chosen to minimize the influence of guy wires and guy rings. For example, for the NRG Systems 60-m TallTower, the measurement heights might typically be as follows:

- *57.2 m*. Sufficiently far below the top of the tower to avoid 3D flow effects; the tower is 203 mm (8 in) wide at this point, so this satisfies the 10-tower-width condition.
- *47.4 m*. Just above the second highest guy ring. Placing the anemometers above the guy ring assures that the wind speed measurements are not affected by the guy wires.
- *32.0 m*. Just above the guy ring at the joint where the tower diameter changes from 254 (10 in) to 203 mm (8 in). This ensures the tower diameter is the same as for the anemometers mounted above, which reduces errors in wind shear.

Lattice Towers. Lattice tower designs vary widely. In addition to the height, important characteristics to know include the face width, compass orientations of the three or four faces (which determine the possible boom directions), heights of guy rings, the diameter of the lattice tubing (or rod), and for preexisting towers, the heights and approximate sizes of other instruments already mounted on the tower. Although it is impossible to define suitable anemometer heights for every situation, typical heights for new, dedicated lattice towers are listed below.

- *80 m*. This height represents the approximate hub height of typical utility-scale wind turbines. The tower top should be at least 10 face widths above this height so that wind flow over the top of the tower does not affect the speed readings.
- *60 m*. This intermediate measurement level is included to provide redundancy and to help define the shear profile. The height chosen should be such that measurements are not affected by guy wires. The effect of guy wires can persist for surprisingly long distances—as far as 40 wire diameters (about 0.25 m for a 6-mm wire) downstream.
- *40 m*. This is approximately the minimum height reached by the blade tip on a large wind turbine.

Mounting. Most anemometers are mounted on booms extending horizontally from the side of the tower. The booms should be made of hollow tubing rather than solid stock, and care should be taken that sensor drainage holes are not blocked.

To produce the most accurate estimate of wind shear, it is strongly recommended that the booms at different heights point in the same direction so that the influence of the tower on the speed readings at each height will be similar. For the same reason, the topmost anemometers should be at least 10 tower widths below the top of the tower; placing them too close to the top can result in a significant error in the estimated wind shear.

Of course, all booms should be as level as possible for an accurate horizontal speed reading. Equally important, they should be long enough so that the influence of the tower on the measured speeds (aside from when the anemometer is directly shadowed by the tower) is small. A boom that is too short can result in significant errors in wind speed measurements (for example, a slowdown when the anemometer is upwind of the tower or a speedup when it is off to the side). According to specifications published by the IEC for wind turbine power performance testing,[1] anemometers should be separated by at least 7 tower diameters from the tower. For example, if the tower diameter is 203 mm (8 in), the minimum recommended distance is 1.4 m (4.7 ft). For lattice towers of relatively low porosity, the distance can be reduced to perhaps 3.75 tower widths (for triangular lattice towers, the tower width is the width of one face).

Anemometers should be mounted well above the booms to minimize their influence on the measured speed. According to the same IEC specifications, the minimum vertical separation between the boom and anemometer is 7 boom diameters, although a much larger separation (12 or 15 diameters) is the norm (for rectangular stock, the diameter is the height of the vertical face).

At two or more heights, and usually for the top two heights, the anemometers should be deployed in pairs on separate booms. This redundancy reduces data losses caused by sensor failures and tower shadow. A typical configuration on tubular towers is for the boom pairs to be oriented 90° from one another and 45° on either side of, and facing into, the prevailing wind direction. On triangular lattice towers, the booms are usually mounted on two tower faces 120° apart and 60° on either side of the prevailing wind direction. However, it may be advisable to depart from these guidelines if there are strong secondary wind directions. For example, if the wind commonly comes from both the east and west, it may be best to mount the anemometers toward the north and south, 180° apart, if possible. The charts in Figure 5-7, which show typical patterns of wind flow disturbance around tubular and triangular lattice towers, may be used for guidance in boom placement.

Mounting an anemometer on a vertical boom at least 7 tower diameters above the tower top is a good configuration for obtaining accurate speed measurements largely unaffected by the tower in all directions. A "goal post" configuration with two such anemometers providing redundancy is also in common use, especially for wind turbine power performance testing. Readings from such vertically mounted anemometers should not be combined with readings from horizontally mounted anemometers to estimate shear, for the familiar reason that the differing effects of the tower on the wind flow could cause significant errors. For this reason, a vertical top-mounted anemometer does not eliminate the need for a pair of horizontal side-mounted anemometers near the top of the tower.

[1] The authors thank the IEC for permission to reproduce information from its International Publication IEC 61 400-12-1 ed.1.0 (2005). All such extracts are copyright of IEC, Geneva, Switzerland. All rights reserved. Further information on the IEC is available from www.iec.ch. The IEC has no responsibility for the placement and context in which the extracts and contents are reproduced by the author, nor is IEC in any way responsible for the other content or accuracy therein. IEC 61 400-12-1 ed.1.0 Copyright © 2005 IEC Geneva, Switzerland. www.iec.ch.

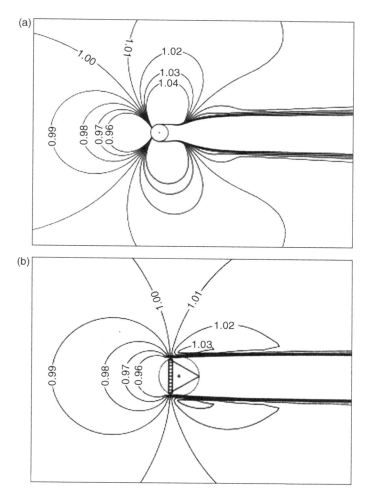

<u>Figure 5-7,</u> Illustration of how the airflow is distorted close to the tower: (a) a tubular tower and (b) a lattice tower. The wind approaches from the left side of each image. The lines represent contours of constant ratio of the disturbed wind speed to the free-stream wind speed. *Source:* IEC 61 400-12, Annex G–Mounting of Instruments on the Meteorology Mast.

5.7.2   Wind Vanes

Wind vanes are generally deployed at two heights. It is customary to install them on booms that are at least 1 m below the nearest anemometer booms to avoid interference with the speed measurements. If it is not practical to mount a vane on its own boom, it should be placed on the anemometer boom about halfway between the anemometer and the tower face. This ensures that the vane disturbs the anemometer readings only when the anemometer is already in the tower's shadow. In addition, it is recommended that the vanes be oriented at least $10°$ away from any guy wire to avoid interference with the vane's rotation as the wires slacken between maintenance visits.

Care must be taken with direction vanes to ensure an accurate direction reading relative to true or magnetic north. Ideally, the wind vane deadband should be oriented along the boom toward the tower. This not only ensures that the vane does not spend a great deal of time in the deadband but also allows the deadband orientation to be easily verified from the ground with a sighting compass or submeter GPS. The deadband orientation must be documented and entered in the data logger software for the logger to correct and report the wind direction relative to true or magnetic north. Consult the sensor or logger manufacturer's recommendations for determining and reporting the deadband position.

5.7.3   Temperature and Other Sensors

A shielded temperature sensor should be mounted on a horizontal boom at least 1 tower diameter from the tower face to minimize the tower's influence on air temperature. The sensor should be well exposed to the prevailing winds to ensure adequate ventilation at most times. When a set of paired temperature sensors is used for ΔT measurements (Section 4.2.3), both sensors should be oriented in the same manner (at different heights) to ensure that they are exposed to similar conditions. If possible, mount the sensors on the northern side of the tower (southern side in the southern hemisphere) to limit heating from direct solar gain; this configuration also reduces the influence of thermal radiation from the tower's surface.

Other sensors should be mounted on the mast according to the supplier's instructions.

5.7.4   Data Loggers and Associated Hardware

Data loggers should be housed along with their cabling connections, telecommunications equipment, and other sensitive components in a weather-resistant and secure enclosure. One can usually be purchased from the data logger supplier. Desiccant packs (usually provided with the logger) should be placed in the enclosure to absorb moisture, and all openings, such as knockouts, should be sealed to prevent damage from precipitation, insects, and rodents. It is also important that all cabling that enters the equipment enclosure have drip loops to prevent rainwater from flowing down the cable to the terminal strip connections, where moisture can cause corrosion.

The enclosure should be mounted on the tower at a sufficient height above ground to be beyond the likely maximum snow depth for the site. Where applicable, the cellular communication antenna should be attached at an accessible height, usually right above the data logger enclosure. If a solar power system is being used, the solar panel should be placed above the logger enclosure to avoid shading and should face south (north in the southern hemisphere) at an angle that will produce sufficient power during the winter, when the sun's apex is low. A near-vertical orientation may be desirable to minimize dust and dirt buildup, which can reduce output.

5.7.5   Sensor Connections and Cabling

The manufacturer's instructions for sensor and data logger wiring configurations should be followed. General guidelines include the following.

- Exposed sensor terminal connections should be sealed with silicone caulking and protected from direct exposure with rubber or plastic boots.
- Sensor wires along the length of the tower should be wrapped and secured with UV- and exposure-resistant wire ties or electrical tape. All slack should be removed as the sensor wires are wrapped around the tower or tower leg. Excessive slack can allow the sensor wires to move in the wind, eventually causing them to break.
- If not installed by the manufacturer, consider installing metal oxide varistors (MOVs) across each anemometer's and wind vane's terminals for added electrical transient protection.
- Where chafing can occur between the sensor wires and supports (such as tilt-up tower anchor collars), the wires should be protected and secured appropriately.

5.7.6   Grounding and Lightning Protection

Grounding equipment is especially important for modern electronic data loggers and sensors, which can easily be damaged by electrical surges caused by electrostatic discharge, lightning, or a difference in ground potential.[2] Most tower and data logger manufacturers provide grounding kits. However, different monitoring areas may have different requirements. Sites prone to lightning activity require an especially high level of protection. Additional protective equipment can often be purchased from the data logger manufacturer or made from common materials found at a hardware store. As part of the planning process, the frequency of lightning activity at the site should be investigated. References such as the lightening frequency map shown in Figure 5-8, as well as meteorological agencies, can provide useful guidance. Even with complete protection, it cannot be guaranteed that equipment will survive a direct lightning strike.

Basic Guidelines. The single-point grounding system, presented in Figure 5-9, is the recommended configuration. This setup minimizes the potential for developing an offset voltage by a grounding loop. In this system, the down conductor wire (10 gauge or less) is directly connected to earth ground via a grounding rod, buried ring, or plate (or a combination of these). It should not be routed through the data logger's grounding stud. The sensor drain or shield wires are electrically tied to the same earth ground via the data logger's common grounding bus (terminal strip). Earth ground is

---

[2]For further information regarding grounding, refer to the appropriate electric code for the country or region in question (e.g., in the United States, National Electrical Code: Article 250–Grounding and Bonding; in Europe, IEC 60 364; in Canada, Canadian Electrical Code; in the United Kingdom, BS 7671; and in France, NF C 15–100).

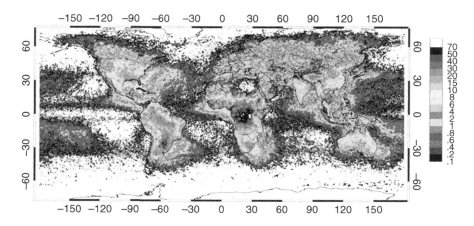

<u>Figure 5-8.</u> Global map of annual lightning strike frequency (in flashes/km^2/ year). *Source:* US National Aeronautics and Space Administration.

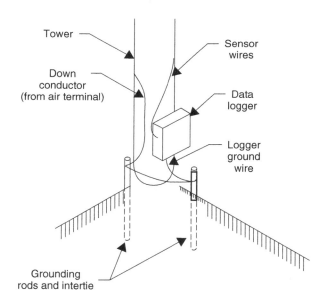

<u>Figure 5-9.</u> Single-point grounding system.

an electrical potential (voltage) level referenced to the earth. Typically, the grounding rod, ring, and plates are copper based to provide a low resistivity for charge dissipation.

The dimensions of the grounding instrument will determine the contact surface area with the soil, a key element for proper system grounding. A combination of grounding instruments (grounding rod, buried ring or plate) can be used to enhance the contact area if they are all electrically connected. At least one 12.5-cm ($\frac{1}{2}$-in)-diameter, 2.4-m (8-ft)-long grounding rod is needed to provide an adequate soil contact area. A deoxidation agent should be applied to all mechanical grounding connections to ensure low resistance to ground. The grounding rods should be free of nonconducting coatings, such as paint or enamel, which can interfere with a good soil contact. All grounding rods must be driven below surface. Where rock is encountered, the rod can be driven in at a 45° angle or buried in a trench at least 0.6 m (2 ft) deep (the deeper the better). Lastly, all grounding rods must be wired together to provide electrical continuity. The above-soil ends of the rods and their electrical conductor attachments should be protected from damage.

It is helpful to know the resistivity of the soil to select the proper grounding system. This is the electrical resistance to current flow within a unit volume of soil, usually located near the earth's surface. It can be approximated by measuring with a multimeter the resistance between two conductive rods driven into the soil to a specified depth and distance apart. The resistance between the grounding system and the earth should be less than 100 ohm. In general, the lower the resistivity of the earth, the better the earth ground it will provide. Soils with low resistivity (e.g., moist dirt) quickly dissipate any voltage potential that develops between two points and provide a better earth ground. High resistivity soil (e.g., dry sand) can build up a large potential voltage or current that may be destructive. If the resistivity is high, several grounding rods may be required. Where the soil can freeze, grounding rods should be driven below the frost line.

Soil resistivity often changes seasonally. The value in early spring, following a winter thaw, may not reflect the soil conditions during the midsummer lightning season. In addition, towers in arid climates may be prone to electrostatic discharge if system grounding is poorly done. When in doubt, take the conservative approach and provide added protection. It is, in the long run, the least costly route.

On existing towers, the tower's grounding system should be evaluated. If it is deemed adequate, the data logger's ground may be connected to it. If not, a separate earth grounding system should be installed and then physically tied into the existing ground system.

Data Logger and Sensor Grounding. Lightning protection devices, such as spark gaps, transorbs, and MOVs, should be incorporated into the data logging system electronics to supplement grounding. Anemometers and wind vanes are available with MOVs as part of their circuitry, or can usually be outfitted with them. Their primary purpose is to limit the peak surge voltage allowed to reach the protected equipment while diverting most of the destructive surge current. The protection offered for each data logger should be verified with the manufacturer. Additional protection equipment may be needed in lightning-prone areas.

Tower Grounding. Lightning protection equipment must be installed on the tower and connected to the common ground. An example of a lightning protection kit consists of an air terminal installed above the tower top, sometimes referred to as a *lightning rod*, along with a long length of heavy-gauge (10 gauge or less), noninsulated copper wire referred to as the *down conductor* tied to the earth ground (a grounding rod or buried loop). The lightning rod generally is considered in most countries to be part of the tower's height, and therefore, the top of the rod cannot exceed height restrictions.

Additional Transient Protection Measures. To provide extra protection against electrical transients, a number of additional steps can be taken:

- The sensor wires can be connected to an additional bank of spark gaps (or surge arrestors) before they are connected to the data logger input terminals.
- A longer air terminal rod with multipoint brush head may provide protection for side-mounted sensors near the tower top by placing them within the theoretical 45° "cone of protection." The purpose of the air terminal is to provide a low impedance path for streaming away charged particles; the cone of protection is the region below the air terminal where lightning flashes are less likely to occur.
- Longer grounding rods may be used. They provide two benefits: first, the soil's conducting properties generally improve with increasing depth, and second, additional contact surface area is gained. Rods that fit together to reach greater soil depths are available for purchase.
- High compression, welded or copper-clad fittings can be employed for all conductor-rod connections.
- The current-carrying capacity of the down conductor can be increased by increasing the cross-sectional area of the wire (i.e., by reducing the wire's gauge).
- The down conductor can be secured to the tower's metal surface with band clamps (one per tower section). Deoxidizing gel helps ensure a good connection.
- A buried copper ground plate or ground ring can be employed at recommended depths to increase soil contact area. It should be connected with other grounding rods.
- Horizontally mounted air terminals can be installed at various levels and directions on the tower to provide additional points for charge dissipation. Each rod should be tied to the down conductor to avoid affecting sensor readings.
- If the tower is secured with concrete or coated (corrosion-resistant) guy anchors, which do not provide a low resistance path to ground, it is recommended that the guy cables be grounded.

5.8   SITE COMMISSIONING

All equipment should be tested to be sure it is operating before a tilt-up tower is raised or while tower climbing personnel are still aloft. These functional tests should be repeated once the installation is complete. Having spare equipment on hand makes repairs easier if problems are found during these tests. Recommended tests include the following:

- Ensure that all sensors are reporting reasonable values.

- Verify that all system power sources are operating.
- Verify required data logger programming inputs, including site identification number, date, time, sensor slope and offset values, and deadband orientations.
- Verify the data retrieval process. For cellular phone systems, perform a successful data download with the home base computer and compare transmitted values to on-site readings.
- Ensure that the data logger is in the proper long-term power mode.

On leaving the site, the crew should secure the equipment enclosure with a padlock and document the departure time and all other pertinent observations.

5.9 DOCUMENTATION

A complete and detailed record of all site characteristics as well as data logger, sensor, and support hardware information should be maintained in a Site Information Log. An example is provided at the end of this chapter. The following main elements should be included in the log:

- *Site Description.* This should include a unique site designation number, the elevation of the site, the latitude and longitude of the mast and anchors, the installation date, and the commissioning time. The coordinates of the site should be determined at installation using a GPS. Typically, coordinates should be expressed to an accuracy of less than 0.01 min (about 10 m) in latitude and longitude and 10 m in elevation. The GPS readings should be cross-checked by comparing with coordinates obtained from a topographic map, and any significant discrepancies should be resolved.
- *Site Equipment List.* For all equipment (data logger, sensors, and support hardware), the manufacturer, model, and serial numbers; the mounting height and directional orientation (including direction of deadbands, cellular antenna, and solar panel); sensor slope and offset values entered in the logger software; and data logger terminal number connections should be recorded.
- *Telecommunication Information.* All pertinent cellular phone or satellite link programming information should be documented.
- *Contact Information.* All relevant landowner and cellular/satellite phone company contact information should be listed.

5.10 COST AND LABOR ESTIMATES

This section describes the main cost elements to be considered when creating a wind monitoring program budget. The quoted costs are appropriate for North America. In other regions, labor costs may be different (either higher or lower). If equipment is to be imported, import duties should be taken into account. In addition, local manufacturing of some equipment, such as towers, may be a cost-effective option.

The estimated total cost, including equipment and labor, to purchase and install a single 50- or 60-m monitoring station and operate it for 2 years is roughly US $50,000–70,000, not including administrative expenses. For an 80-m lattice tower, the cost range is approximately $100,000–200,000. The actual cost will depend on the specific tower type and selected equipment, site access, proximity to operation and maintenance staff, and the number of site visits required.

The monitoring cost can be divided into three main categories.

- *Labor.* Table 5-1 lists the main tasks to be accounted for when budgeting for labor. Some tasks require just one person, others, especially those that deal with equipment installation and maintenance, require two or more.
- *Equipment.* Equipment costs can be obtained from vendors once the measurement specifications are determined. Other items to consider in the budget are shipping charges, taxes, import duties, insurance, spare parts, and the tools needed to install and service the tower. The estimated total equipment cost for a single site that uses a 60-m tilt-up tubular guyed tower equipped with three levels of sensors is typically $15,000–20,000.

Table 5-1. Labor tasks to account for when budgeting

Administration
Program oversight
Measurement plan development
Quality assurance plan development

Site selection
In-house remote screening
Field survey and landowner contacts
Obtain land use agreement and permit

Equipment
Specify and procure
Test and prepare for field
Installation (typically four people)

Operation and maintenance
Routine site visits (one person)
Unscheduled site visits (two people)
Preventative maintenance activities
Calibration at end of period
Site decommissioning (four people)

Data handling and reporting
Validation, processing, and report generation
Data and quality assurance reporting

- *Expenses.* Related expenses include travel, land lease fees, cellular or satellite phone fee (if applicable), and sensor calibration. Travel costs should account for the anticipated number of field trips required to select, install, maintain, and decommission a site. Some field trips may require overnight lodging and meals. Remote data transfer using a cellular or satellite phone link can add costs, typically $50–70 per month, depending on the number and duration of calls and the rates.

Economies of scale can be achieved with multiple towers. Most of the savings are in labor. Roughly speaking, labor expenses for each additional site should be about 30% less than those for a single site, depending on the number of sites and their proximity to one another. Travel expenses can be reduced if more than one site is visited in a single trip. Savings on equipment can be realized through vendor discounts and by sharing installation equipment (e.g., gin pole, winch kit) among sites. Overall, the total cost to install and operate a second site is typically about 10–15% less than the cost for the first site. The average cost per tower for a five-tower monitoring network is about 20% less than that of a single tower.

A resource assessment program should have a project manager, a field manager, and a data manager, plus additional support staff such as field technicians. Their roles are defined below. Some staff may be able to perform multiple roles.

- The project manager directs the wind monitoring program and ensures that human and material resources are available in a timely manner to meet the program's objectives. The project manager should also oversee the design of and adherence to the measurement and quality assurance plans.
- The field manager is responsible for installing and maintaining the monitoring equipment and transferring the data to the home office. This person, or an assistant, should be available to promptly service a site whenever a problem arises. The installation and decommissioning of tilt-up met towers, as well as service visits that require towers to be lowered, necessitate a crew of at least four people. For lattice towers, which can be serviced while upright by tower climbers, it is recommended that at least two tower climbers be present during maintenance.
- The data manager is responsible for all data-related activities, including data validation and report generation. Familiarity with meteorology and the monitoring site and equipment and a close working dialog with the field manager are essential to properly validate and interpret the data.
- The field technicians work closely with the field manager to organize and coordinate many aspects of the resource assessment campaign. The technicians are responsible for evaluating and siting meteorological towers, procuring hardware, overseeing tower installations, conducting verifications of existing towers, and providing maintenance support during campaigns.

5.11   QUESTIONS FOR DISCUSSION

1. Suppose you are leading a wind resource assessment campaign for a developer interested in building a 50-MW wind project. Create a set of specifications for an instrumentation package and tower. Indicate your desired tower height; the heights of the speed, direction, and temperature sensors as well as any other sensors you may choose; and the number of sensors at each height.

2. Use the Internet to obtain price and delivery quotes for your preferred package from at least two vendors. Make sure you take into account taxes and import duties. What is the total cost of each package? Is the equipment (including tower) to be provided substantially the same, or are there differences? In your opinion, which meets your employer's needs in the most cost-effective manner?

3. Describe how you would go about acceptance testing your preferred instrument package once it is received, as well as field preparation.

4. Identify and describe at least four specific safety guidelines covered in this chapter. State why they are important and what could happen if they were not followed.

5. Suppose you are on a site and your GPS is broken. Given a magnetic compass, how would you determine true north? What is the approximate magnetic declination in your area?

6. You are in charge of installing a 60-m tilt-up tubular tower in a farmer's field. The manufacturer has shipped the tower with anchors that are 44 mm wide and 381 mm long. Should you be concerned? Why? If so, what should you do?

7. Explain the function of a gin pole on a tilt-up tower. Suppose you are installing a tilt-up tower, and you find the wind is moderate and from the west. In which direction should the gin pole face, and why?

8. You have been asked to assess an instrumented tower installed by a company whose wind project your employer is interested in investing in. You arrive at the site, and using a laser range finder, you find that it is a 60-m-tall, 0.5-m-wide, custom-built square lattice tower. You further observe a single anemometer mounted on a vertical boom extending about 0.5 m above the top, and a single anemometer mounted on a horizontal boom extending 2 m from the side at a height of about 45 m. There is also a direction vane on the same boom. You advise the company that this monitoring setup is inadequate. Why? How would you advise that it be improved?

9. Explain the rationale for installing two anemometers at the same height. What is the purpose of setting the boom orientations 90° apart and 45° on either side of the prevailing wind direction? Why should booms at different heights be oriented in the same directions?

10. What is the purpose of grounding the equipment on a met mast? In the region where you live, what code governs electrical installations, including grounding?

11. Using the Internet, investigate available sources for information on lightning strikes in an area you might consider for building a wind project. Is the risk high or low? At what time of year is the risk highest?

12. Explain why it is important to verify and document the installation of a met tower and its monitoring equipment.

SUGGESTIONS FOR FURTHER READING

NRG Systems, Inc. 60 m and 50 m XHD TallTower installation manual and specifications, April 2010. Available at http://www.nrgsystems.com/FileLibrary/03e8c4ba85de4996a33 abe3227d57873/NRG%2060m%20and%2050m%20XHD%20TallTower%20Installation% 20Manual%20-%20Rev%202.01.pdf.

International Electrotechnical Commission (IEC) 61400-12, Wind Turbine Power Performance Testing. Annex G—mounting of instruments on the meteorology mast. Available at http://webstore.iec.ch/webstore/webstore.nsf/mysearchajax?Openform&key=61400&sort ing=&start=1&onglet=1. (Accessed 2012).

International Energy Agency. Program for Research and Development on Wind Energy Conversion Systems. Expert Group Study on Recommended Practices for Wind Turbine Testing and Evaluation: Topic 11. Wind Speed Measurement and Use of Cup Anemometry', Second Print, 2003. p. 60. Available at http://www.ieawind.org/Task_11/Recommended_Pract/11%20Anemometry_secondPrint.pdf. (Accessed 2012).

SAMPLE SITE INFORMATION LOG

Form Revision Date:

Site Description	
Site Designation	
Location	
Elevation	
Installation/ Commission Date	
Commission Time	
Soil Type	
Surroundings Description	
Prevailing Wind Direction	
Declination	

Site Equipment List						
Equipment Description	Mounting Height	Serial Number	Sensor Slope	Sensor Offset	Logger Terminal Number	Boom Direction (vane deadband)

Telecommunication Information	
Device Manufacturer	
Device Model	
Device SN	
Network ID	
Phone Number	
Programmer	
Date Programmed	
e-mail Address	
Subject Line	
Password	
Antenna Type	
Antenna Location	
Power Source	

Contact Information	
Landowner Name	
• Address	
• Phone Number	
Cellular/Satellite Company	
• Phone Number	
• Contact Person	
• Contact Extension	

6

STATION OPERATION AND MAINTENANCE

The goal of the operation and maintenance phase of a wind monitoring campaign is to ensure that accurate wind resource data are collected in a reliable manner. A host of problems can occur, causing data losses or erroneous readings. Meteorological instruments can be damaged, their mountings can slip, and towers can bend or fall. In addition, various system components from sensors to guy wires may require periodic maintenance to function properly.

To address these needs, a simple but thorough operation and maintenance plan should be developed and implemented. Key elements of the plan include scheduled and unscheduled site visits; inspection procedures, checklists, and logs; calibration checks; and spare parts. Guidelines to develop such a program are provided in this chapter.

Although a sound operation and maintenance plan is critical, ultimately, the success of the program depends on the dedication and training of the field personnel. They should be briefed on all aspects of the program and have a working knowledge of the monitoring equipment. They should also be conscientious and detail oriented, observant note takers, and good problem solvers.

Wind Resource Assessment: A Practical Guide to Developing a Wind Project, First Edition.
Michael Brower et al.
© 2012 John Wiley & Sons, Inc. Published 2012 by John Wiley & Sons, Inc.

6.1   SITE VISITS

Site visits are either scheduled or unscheduled. A scheduled visit normally serves one of two purposes. The first is to carry out routine inspections and maintenance tasks, such as replacing batteries, adjusting guy wire tensions, and checking boom orientations. This type of visit is typically done once every several months, depending on the expected life of the equipment and manufacturers' guidelines. The second purpose, which only applies if remote options are not available, is to retrieve data manually from the data logger. The frequency of visits for manual data retrieval depends on the storage capacity of the logger but in any event should be no less often than once every 2 weeks. This schedule helps ensure that sensor problems are promptly detected and fixed so that data losses can be kept to a minimum (the data retrieval process is detailed in Chapter 7).

An unscheduled site visit is warranted whenever a significant problem with the data collection is suspected. This could happen, for example, because of a failure to retrieve data remotely, nonsensical sensor readings, a marked change in the readings of one or more sensors relative to others, and other issues. Reports of severe weather, including high winds, lightning, or icing conditions, around the site may also indicate that a visit is warranted. In such cases, an unscheduled visit should be made as soon as possible to minimize the potential loss of data. Both the program budget and staffing plans should anticipate at least one unscheduled visit to each tower every year.

6.2   OPERATION AND MAINTENANCE PROCEDURES

The operation and maintenance program should be documented in an Operation and Maintenance Manual. The purpose of this manual is to provide field personnel with comprehensive, step-by-step procedures for carrying out both scheduled and unscheduled operation and maintenance tasks. The manual should include the following elements.

6.2.1   Project Description and Operation and Maintenance Philosophy

The manual should start by describing the project and its overall goals. The important role of the technician in maintaining data quality and completeness should be emphasized.

6.2.2   System Component Descriptions

The manual should provide a brief description of all instruments (anemometers, wind vanes, temperature probes, data loggers, and others) and how they work. Detailed information, such as manufacturers' manuals, should also be available for reference. Such information helps field technicians to perform their jobs well and respond effectively to unexpected situations.

6.2.3 Routine Instrument Care Instructions

All equipment requiring routine maintenance should be identified and maintenance instructions provided in the manual. Routine maintenance tasks can be divided into two categories: those related to the tower structure, supports, and grounding, and those related to the instruments and data acquisition system. The following sections lists the tasks typically performed in each category.

Tower Structure, Supports, and Grounding

1. *Anchor Condition*
   - Check for signs of rust or damage.
   - Assess movement of the anchors over time.
   - Verify the integrity of the anchor connections; for example, the anchor resistance may have changed if an animal has burrowed near the connection point.

2. *Guy Wire Condition*
   - Check that the guy wires are properly tensioned in accordance with the manufacturer's guidelines. Tension the guy wires if necessary.
   - Inspect the wires and connection points for signs of rust or corrosion.
   - Ensure that the appropriate number of wire clips were used to secure the wires and that the clips are in good condition.

3. *Tower Condition*
   - Check for signs of rust or damage.
   - Confirm that the tower is plumb and straight.
   - For tubular towers, examine the tower for signs of flaring at the connection points between tower sections.
   - Inspect the baseplate or foundation to ensure that it is not sinking or distorted and is otherwise free from damage.

4. *Grounding System*
   - Verify that the grounding system is connected properly and the electrical contacts are in good condition.

Instruments and Data Acquisition System

1. *Sensors*
   - Inspect the sensor booms and instrument pillars to make sure they are level or plumb and in good condition.
   - Booms are known to slip or rotate; confirm that all booms are at the correct heights and point in the correct directions.
   - Replace any sensors that data analysis shows may be deteriorating or may have failed (Chapter 9).

- Replace wind vanes and anemometers regularly as part of a preventive maintenance plan. A typical replacement schedule is once every two years. To ensure continuity of the data record, replace only one of a pair of sensors at the same height in a single visit.
- If any doubts arise concerning the stability of an anemometer's calibration, it may be advisable to have the anemometer removed and recalibrated. Comparing the pre- and postdeployment calibration coefficients allows any significant drift in the anemometer response to be detected.
- Some anemometer types require periodic refurbishment such as ball bearing replacement and recalibration.

2. *Data Acquisition System*
   - Inspect the logger and the enclosure for signs of corrosion, damage, moisture, and the presence of rodents or insects.
   - Check the wiring panel to ensure a good connection to the sensors.
   - Check battery voltage and replace batteries as needed.
   - If the batteries are charged by a solar system, the solar panels should be cleaned and their alignment checked and the panels, wiring, and electrical connections should be examined for cracks and water resistance.
   - If a diesel or other generator is used, it should be tested and refueled.

6.2.4 Site Visit Procedures

All site visits should be carried out in three stages: in-house preparation, on-site tasks, and site departure procedures.

1. *In-House Preparation*
   - The field team should be informed of the reasons for the visit and the tasks that need to be carried out. For example, is this a scheduled inspection and maintenance visit or is it in response to a possible problem with the tower or equipment? Will it be necessary to lower the tower to the ground or climb it, or can the tasks be performed with access only to the equipment at the tower's base?
   - Where appropriate, the landowner should be informed of the visit in advance. Aside from being courteous, maintaining good relationships with landowners can help the wind project succeed over the long run.
   - The team should be provided with a complete set of tools, supplies, equipment manuals, and spare parts to accomplish all tasks. The Site Visit Checklist (Section 6.3) specifies the required tools and supplies. This list should include all equipment necessary to download the site data, such as laptop computers with associated cables and special hardware.
   - An extra memory card is a must, and it should be tested in-house before being taken into the field. This is especially important when swapping memory cards

is the main method of data retrieval. The testing may require a spare in-house data logger.

- It is important to ensure that the number of personnel being sent to the site is adequate for the tasks to be performed. For safety, tower climbing requires at least two people—one to climb and the other to be ready to respond to problems from the ground. Raising and lowering tilt-up towers usually requires at least four people.
- The team leader should inform the home office of where the team will be and when it is likely to return. For sites in especially remote or dangerous areas, the team should stay in regular contact with the home office, and management should be promptly notified if contact is lost for an extended period or if the team fails to return on time.

2. *On-Site Procedures*

- It is recommended that on-site work begin with an informal meeting in which the day's plans are reviewed. This also offers an opportunity to verify compliance with any personal protective equipment (PPE) requirements and safety procedures.
- If data are to be retrieved during the visit, this should be done first to minimize the risk of data loss from operator error, static discharges, or electrical surges during handling or checking of other system components.
- No matter the purpose, each visit should include a thorough visual inspection (with binoculars or digital camera), as well as testing, when applicable, to detect damaged or faulty components. The results should be recorded on the Site Visit Checklist. The inspection should include the following:
  - Data logger
  - Sensors
  - Communication system
  - Grounding system
  - Wiring and connections
  - Power supply (or supplies)
  - Support booms
  - Tower components (for guyed tower systems this includes anchors, guy wire tension, and tower vertical orientation).
- Scheduled component replacements (e.g., batteries), operational checks, and troubleshooting should be carried out during the visit if necessary. Troubleshooting guidelines should be developed before the first site visit.
- The instantaneous data logger readings should be examined to verify that all measured values are reasonable.
- The Site Visit Checklist should be filled out to ensure that all operation and maintenance tasks have been completed and the necessary information documented.

3. *Site Departure Procedures*

- The correct functioning of the data retrieval system should be verified before leaving the site. This involves completing a successful data transfer with the home-base computer (for remote systems) or in-field laptop computer (for manual systems). For remote systems, data transfer can be verified at the site through the use of a data drop box (such as an email account or FTP folder) that can be accessed from the field. This simple test will ensure that the system is operating properly and the remote communication system (antenna direction and phone connections) was not inadvertently altered during the visit.

- The data logger should be returned to the proper long-term system power mode. Some models have a low power mode for normal operation to conserve system power. Neglecting to invoke this mode will significantly reduce battery life and may cause data loss.

- The data logger enclosure should be secured with a strong, high quality padlock to discourage curious visitors and vandalism.

- The departure time should be recorded and all work performed and observations made should be entered on the Site Visit Checklist.

6.3 DOCUMENTATION

The Site Visit Checklist, which follows the procedures outlined in the Operation and Maintenance Manual, is a helpful tool for the field technician. It provides a reminder of what needs to be done on each visit and serves as a record of the actions taken. An exact, detailed record can help to explain periods of questionable data and may prevent a significant amount of data from being discarded during data validation.

For these reasons, a standardized checklist should be developed, completed for each site visit, and kept on file. Example information and activities to detail in the checklist include the following:

1. *General Information.* Site name, technician name(s), date and time of site visit, and work to be performed. (Note: To avoid confusion, all times recorded here and elsewhere should be in local standard time, or LST.)

2. *In-House Preparation.* List of necessary tools, equipment and supplies (including spares), documentation, maps, and safety items.

3. *On-site Activities.* A sequential list of the various site activities including equipment checks, data retrieval, tower-related work (raising and lowering procedures), and departure activities.

4. *Findings and Recommendations.* A detailed account of the work performed, findings, and observations, and if applicable, further recommended actions.

A sample Site Visit Checklist is provided at the end of this chapter.

6.4   SPARE PARTS INVENTORY

The operation and maintenance plan must anticipate equipment malfunctions and damage. To minimize downtime, an adequate spare parts inventory should be maintained and available for use during site visits. The inventory should include replacement up-tower items such as sensors, booms, and associated mounting hardware. Additional items may be needed. The following points should be considered when determining inventory needs:

1. *Size of the Monitoring Network.* The size of the spare parts inventory depends in part on the number of towers in the monitoring network. As a guide, a network with six monitoring towers should have a parts inventory sufficient to outfit two towers. For networks of this size and larger, it is also advisable to have a spare data logger and remote communications device on hand.

2. *Environmental Conditions.* Towers in areas prone to extreme weather should have additional spares. Recommended extras include spare anemometers, wind vanes, and sensor mounting booms.

3. *Equipment Availability.* The inventory of spares should be increased for items that require an extended lead time for delivery from the supplier. The turnaround time for critical items, such as data loggers and sensors, is particularly important.

4. *Operations and Maintenance History.* Inventories should be adjusted during the program based on experience at each site. Sometimes sensors fail more often than expected, so additional spares may be required.

5. *Vandalism.* Certain sites may be prone to vandalism. Cups on anemometers are sometimes targeted for shooting practice, and equipment mounted near the ground, such as solar panels and grounding system, may be stolen. If frequent access is not needed for data retrieval, consider mounting the base equipment (logger and peripherals) higher on the tower, out of easy reach. If vandalism is a concern, consider installing a fence around the base of the tower.

6.5   QUESTIONS FOR DISCUSSION

1. The data recovery is defined as the percentage of all possible records in a given period that are deemed valid. The data loss is 100% minus the data recovery. (i) Explain how the data recovery can be affected by the frequency of data retrieval from a wind monitoring station. (ii) Suppose the logger fails an average of once a year. If the data are retrieved (either manually or remotely) and screened for problems every 2 weeks, and any problems are immediately followed by a site visit, what is the greatest data loss that might result from the failure? What is the corresponding data recovery? (iii) How does the answer change if the data are retrieved once a week? Once a month?

2. Summarize the inspection routine for a scheduled maintenance visit to a guyed tilt-up tower. Include all major components or equipment to be assessed and how the assessment should be done.

3. Discuss three safety precautions mentioned in this chapter. Explain what can happen if these precautions are not respected.

4. Suppose your company has three towers in the field. Develop a specification for the instrumentation on each tower (e.g., numbers and types of anemometers, wind vanes, and other sensors), and define and price—using the Internet—an adequate spare parts inventory.

5. Why might it be a good idea from time to time to remove the anemometers on a wind monitoring tower and send them to be recalibrated? For one of the towers you specified in Question 4, describe a suitable removal schedule.

Sample Site Visit Checklist

A. General Information

Site Designation	
Site Location	
Crew Members	
Date(s)	
Time (LST)	Arrival:	Departure:
Visit Type (check)	Scheduled ☐	Unscheduled ☐
Work Planned		

B.   In-House Preparation

Check each box to denote the items have been acquired.

☐   In-house support person: _____

☐   Copy of Site Information Log.

☐   Acquire necessary tools, equipment, and supplies.

    ☐   Electrical supplies: voltmeter, fuses, tape, connectors, cable ties, batteries crimpers, etc.

    ☐   Wrenches, pliers, screwdrivers, nut drivers, hex set, sledgehammer, wire cutters, etc.

    ☐   Misc. equipment: silicone, magnetic level, binoculars, camera, GPS, etc.

    ☐   Spare parts: cabling, anchors, booms and mounting hardware, etc.

       1)   Sensors:

          1)   Sensor: _____ Serial # _____ Slope/Offset: _____/_____

          2)   Sensor: _____ Serial # _____ Slope/Offset: _____/_____

          3)   Sensor: _____ Serial # _____ Slope/Offset: _____/_____

       2)   Data logger: Serial # _____

☐   Road and topographic site maps.

☐   Rental equipment: jackhammer w/compressor, truck/trailer, etc.

☐   Winch with 12 - V battery and battery charger.

☐   Gin pole and associated hardware.

☐   Safety equipment: Hard hats, gloves, appropriate clothes, first aid kit, etc.

☐   Manufacturer's manuals for installation and troubleshooting (sensors, data logger, etc.)

Additional Information/Comments: _____

_____

_____

_____

Site Designation: _____

C.   General On-Site Activities

Check the appropriate box. If No, provide and explanation below.
- General Visual Inspection
  Yes ☐ No ☐ Area free of vandalism?
  Yes ☐ No ☐ Tower straight?
  Yes ☐ No ☐ Guy wires taut and properly secured?
  Yes ☐ No ☐ Solar panel clean and properly oriented?
  Yes ☐ No ☐ Wind sensors intact, oriented correctly, and operating?
  Yes ☐ No ☐ Sensors, solar panel, and antenna are free of ice or snow?
  Yes ☐ No ☐ Grounding system intact?
  Yes ☐ No ☐ Cellular antenna correctly orientated?
  Findings/Actions: _____

- **Data Retrieval**
  Manual ☐ Remote ☐ (download method)
  Yes ☐ No ☐ Successful download? If No, provide explanation below.
  Findings/Actions: _____

- **Tower Lowering Activities**
  Yes ☐ No ☐ Check all anchors, no signs of movement?
  Yes ☐ No ☐ Winch secured to anchor and safety line connected to vehicle chassis?
  Yes ☐ No ☐ Gin pole assembled with safety cable and snap links tape?
  Yes ☐ No ☐ Tower base bolt tight?
  Yes ☐ No ☐ Gin pole safety rope attached and tensioned properly (gin pole straight)?
  Yes ☐ No ☐ Weather conditions safe?
  Yes ☐ No ☐ Personnel clear of fall area?
  Yes ☐ No ☐ Note start time of tower lowering. _____(LST)
  Yes ☐ No ☐ Winch battery connected and terminals covered?
  Yes ☐ No ☐ Lifting guy wire attachments to gin pole checked?
  Findings/Actions: _____

- **On-Ground General Activities**
  Yes ☐ No ☐ Sensor and ground wires securely attached?
  Yes ☐ No ☐ Grounding system intact and secure?
  Yes ☐ No ☐ Sensor boom clamps secured?

Site Designation: _____

C. General On-Site Activities (continued)

- **On-Ground General Activities (continued)**

  Yes ☐ No ☐ Boom orientation OK?

  Yes ☐ No ☐ Boom welds OK?

  Yes ☐ No ☐ Vane deadband orientation as reported on Site Information Log?

  Yes ☐ No ☐ Sensors level and oriented correctly?

  Yes ☐ No ☐ Sensor wire connections secure and sealed with silicone?

  Yes ☐ No ☐ Signs of sensor damage?

  Yes ☐ No ☐ Sensor outputs checked and functioning properly?

  Yes ☐ No ☐ Sensor serial numbers as reported on Site information Log?

  Yes ☐ No ☐ Sensor and/or data logger replacement? If Yes:

  1) Sensor: _____ Serial #_____ Slope/Offset: ____/____
     Height: ____ Orientation: ____
  2) Sensor: _____ Serial #_____ Slope/Offset: ____/____
     Height: ____ Orientation: ____

  Findings/Actions: _____

- **Tower Raising Activities**

  Yes ☐ No ☐ Guy wire collars positioned correctly?

  Yes ☐ No ☐ Lifting lines and anchor lines properly attached?

  Yes ☐ No ☐ Gin pole secure, lines tensioned, gin pole straight, snap links taped?

  Yes ☐ No ☐ Weather conditions safe?

  Yes ☐ No ☐ Guys properly tensioned?

  Yes ☐ No ☐ Tower straight?

  - Note online time: _____ (LST)

- **Site Departure Activities**

  Yes ☐ No ☐ Successful data transfer with office computer?

  Yes ☐ No ☐ Checked antenna and phone connections?

  Yes ☐ No ☐ Is data logger data/time correct?

  Yes ☐ No ☐ Secure data logger enclosure with lock?

  Yes ☐ No ☐ Clean area?

  Yes ☐ No ☐ Guy wires clearly marked?

  Findings/Actions: _____

Site Designation: _____

D.   Findings and Recommendations

Yes ☐ No ☐ Further actions required? If Yes, describe below:

7

DATA COLLECTION AND HANDLING

The main objective of the data collection and handling process is to make the meteorological measurements from a wind monitoring campaign available for analysis while protecting them from tampering and loss. This chapter highlights the key aspects of meeting this objective, including data storage, retrieval, protection, and documentation.

7.1 RAW DATA STORAGE

Data are typically stored in the data logger in a compact, binary (non-text) file format, which cannot be read without special software. In this form, they are commonly referred to as *raw data*. To ensure high data recovery during the monitoring program, the data logger's internal storage medium should be nonvolatile, meaning its data are retained even if the logger loses power; and the raw files should be retrieved from the logger before its storage capacity is reached.

Once transferred from the logger to a computer, it is critical that the logger's raw data files be permanently archived and preserved, as they provide the best evidence that the data collected from the mast have not been altered, and they also contain an

Wind Resource Assessment: A Practical Guide to Developing a Wind Project, First Edition.
Michael Brower et al.
© 2012 John Wiley & Sons, Inc. Published 2012 by John Wiley & Sons, Inc.

original record of the conversion constants applied to the raw sensor readings (this is of particular importance for those data loggers that only provide the converted data rather than the raw outputs of the sensors). Without access to these files, an independent reviewer may have doubts about the reliability and accuracy of the measurements.

7.1.1   Data Storage Types

The following is a list of common raw data storage media.

- *Data Card*. These are small, removable storage devices widely used in digital cameras, camcorders, and similar applications under brand names such as Sony Memory Stick, MultiMediaCard (MMC), and SecureDigital (SD) card. Many laptops are equipped to read such data cards directly. The data can also be imported into a laptop or desktop computer through a data card reader.
- *Solid-State Modules (SSMs)*. These nonvolatile devices, also called *internal memory*, are hardwired into the logger. The data are read through a direct cable connection from the logging system to a laptop.
- *EEPROM Data Chip*. This is an older integrated-circuit technology that served as internal memory for earlier loggers. The manufacturer's software and an EEP-ROM reading device are required for data transfer.

A laptop computer is needed if the data are to be transferred on-site. This is the recommended method for manual data retrieval since it allows the integrity of the data to be verified during the site visit. If no laptop is available, then the data card must be replaced with a fresh card and brought back to the company office. Depending on the storage type, special cabling, interface hardware, external power supply, and software may be required, along with portable drives or USB drives.

7.1.2   Data Storage Capacity

The minimum required storage capacity of the logger depends on several factors, including how often the data are retrieved (once every 2 weeks or more often is recommended), the data recording interval (typically 10 min), the number of sensors being monitored (typically 8–12 on a 60-m tower), and the number of parameters calculated and stored by the logger. The capacity of the data storage devices commonly used today, at least 16 MB, is ample for most situations. One exception may be when a data recording interval shorter than 10 min, such as 2 s or 1 min, is desired. Then larger data storage or more frequent data retrieval may be necessary. Another exception may be if the tower is likely to be inaccessible for months at a time (e.g., because of winter snow and ice). Then if the telecommunications uplink fails, the logger may be called upon to store data for up to several months.

Manufacturers usually provide tables or methods to calculate the approximate available storage capacity (in days) for various memory configurations. Capacity estimates should also allow for delays in retrieving the data.

7.2   DATA RETRIEVAL

The selection of a data transfer and handling process (manual or remote) depends on the characteristics of the site and requirements of the monitoring program. In general, remote data retrieval is preferred since it usually involves less staff time and travel expense. However, some sites may not have reliable cellular phone service, and other remote connection options can be expensive. In choosing a method, the following points should be considered:

- labor costs and availability;
- travel time and expense;
- year-round site accessibility;
- availability of cellular phone service;
- communication equipment costs and power needs;
- support systems required (e.g., computers, modems, analysis and presentation software).

7.3   DATA RETRIEVAL FREQUENCY

Whether data are retrieved manually or remotely, frequent retrieval and prompt review are key to achieving high data quality and low losses. A schedule of regular site data transfers, or downloads, should be followed. The maximum recommended interval is 2 weeks. With remote data transfer, the size of the data sets to be transmitted is an important consideration. For reliable transfers, the files should be as small as possible. A weekly schedule may suffice, but a shorter interval, such as every 1–3 days, may be better.

Situations may arise that warrant unscheduled transfers. For example, if sensor irregularities are discovered when the data are reviewed, a follow-up transfer may be called for to see if the problem is persisting. Reports of icing or severe weather near the site could be the cause for making an unscheduled retrieval to make sure the sensors are still working properly. Of course, whenever problems are suspected, a field crew should be dispatched as soon as possible to inspect the tower and instruments.

7.4   DATA PROTECTION AND STORAGE

The following sections offer guidance to minimize the risk of data loss and corruption.

7.4.1   Data Logger

To ensure data are protected while stored in the data logger, proper installation procedures should be followed, including grounding all equipment and using spark gaps.

7.4.2   Computer Hardware

A personal computer is usually the primary location of the working database, so care should be taken to ensure that the computer and especially its hard drives are in good working order and that the data are frequently backed up.

7.4.3   Data Handling Procedures

Improper data handling increases the risk of data loss. All personnel in contact with the data and storage media should be fully trained and should understand the following:

- *Data retrieval software and computer operating system.* Technicians should be aware of all instances in which data can be accidentally overwritten or erased.
- *Good handling practices for all data storage media.* Data cards and hard disk drives should be protected from static charge, magnetic fields, and temperature extremes.
- *Computer operations and safety practices*, including grounding requirements.

To reduce the risk of data loss, the raw logger files should be permanently and safely archived and the working database backed up regularly (at least as often as the data retrieval). Archives and backup copies should be stored in a different location from the main files, and not in the same building. Common data backup methods include the use of CD, DVD, and magnetic tape. Online backup services have recently become popular and are especially secure as well as convenient for frequent backups.

With remote data transfers via e-mail, another very effective data protection strategy is to set up backup e-mail accounts. The e-mailed files can go to different computers in different locations.

7.5   DOCUMENTATION

A Site Data Transmission Report, an example of which is presented at the end of this chapter, should be developed to serve as the master raw data file log for each site. The report can also be used to track the success of remote data transfers and document file backups. The following lists the basic information to include in the Site Data Transmission Report:

- Site designation
- Site location
- Data-transfer method (manual or remote)
- Last transfer date and transfer time (local and GMT)
- Backup system and location
- Transfer interval
- Comments, problems, and actions taken.

This document provides valuable feedback on equipment performance and data completeness. For example, a review of past reports may indicate that although data have been successfully retrieved, establishing and maintaining site communications is becoming increasingly difficult. This may be the first indication of an impending failure of the system's telecommunication system or power supply and suggests that it is time to visit the site before data are lost.

7.6 QUESTIONS FOR DISCUSSION

1. You would like to collect 10-min average and 2-s instantaneous data from the same site. What would be an appropriate storage medium and data retrieval method to allow this?

2. You are installing a measurement tower in a remote area where multiple communication options (satellite, cellular phone) are available. Discuss the costs and benefits of each alternative. Which would you recommend for this area and why?

3. In compiling your routine data transmission report, you notice that a site has not communicated with the base software for 10 days. What potential problems could cause this, which can be diagnosed remotely, and at what point does this necessitate a site visit?

4. You are conducting a site visit to a remote location about which you do not have any information on the data logger configuration. List the equipment you would need to allow for data retrieval from any common logger configuration as well as the appropriate backup equipment you would need on-site.

5. You have a wind monitoring site that can be easily accessed by the landowner for data manual retrieval in all seasons. Discuss the pros and cons of manual versus remote data retrieval, and develop a plan to ensure data integrity in either alternative.

6. Many modern data loggers can be programmed to send data at preset time intervals. Describe the relative merits of different data-transfer intervals, from shorter (daily) to longer (weekly, biweekly, or longer), from both the equipment and general monitoring campaign perspectives.

Sample Site Data Transmission Report   Date of Report

Site Number	Site Name	Comments, Problems, Actions Taken	Site Location	Data Transfer Method	Last Transfer Date	Transfer Time		Backup System/ Location	Transfer Interval
						Local	GMT		

8

GROUND-BASED REMOTE SENSING SYSTEMS

As wind turbines become larger and the size and complexity of wind projects increases, there is a need for wind resource data from greater heights and in more locations across a project area. Ground-based remote sensing, which includes sodar and lidar, can help meet this need. Sodar and lidar measure the wind resource to heights of 150 m or more above ground, well beyond the reach of tilt-up towers. In settings where fixed masts are prohibitively expensive or not technically feasible, sodar and lidar may be the sole source of wind measurements. More frequently, they are used in conjunction with fixed masts, which remain the standard for resource assessment. Although the practice of relying entirely on remotely sensed data remains rare, it is likely to become more common as the cost of the technology decreases, its accuracy and reliability continue to improve, and experience with it grows.

The main advantage of remote sensing is the ability to measure wind characteristics above the heights of most wind monitoring towers and across the rotor plane of modern, large wind turbines. This can reduce the uncertainty in wind shear and energy production estimates. Information on turbulence, vertical motions, and directional shear (veer)—all of which can affect turbine performance—can also be obtained.

Wind Resource Assessment: A Practical Guide to Developing a Wind Project, First Edition.
Michael Brower et al.
© 2012 John Wiley & Sons, Inc. Published 2012 by John Wiley & Sons, Inc.

Another advantage of remote sensing is that the devices can be deployed and moved relatively easily, so that the wind resource can be sampled at a number of locations within a project area, often at less cost and in less time than with tall towers. In some cases, they can be deployed at sites where it is impractical or prohibited to erect towers. Typically, the period of measurement, when the systems are paired with long-term meteorological towers, is from a few weeks to a few months, or however long is deemed adequate to obtain a statistically representative sample of atmospheric conditions.

Both sodar and lidar measure the wind very differently from conventional anemometry. The differences between these measurement systems must be considered when comparing wind characteristics derived from them. One difference is that they measure the wind speed within a volume of air rather than at a point. Another is that they record a vector average speed rather than a scalar average speed. Remote sensing units also behave differently from anemometers under precipitation, in turbulence, and where vertical winds are significant, and their performance can be affected by variations in temperature, complex terrain, and other factors.

The next two sections discuss current industry-accepted practices and techniques for integrating sodar and lidar into wind resource assessment programs. Areas of active research and development are also mentioned. Some providers of remote sensing equipment are listed in Appendix A.

8.1 SODAR

Sodar operates by emitting acoustic pulses (audible chirps or beeps) upward and listening for the backscattered echoes. The scattering is caused by turbulent eddies (small-scale fluctuations in air density) carried along by the wind. The movement of these eddies causes a Doppler frequency shift—the same effect that makes an ambulance siren seem to change pitch as the ambulance approaches and passes an observer. The frequency shift is analyzed by software, which determines the radial wind velocity along the transmitted pulse; the horizontal and vertical wind velocities are then derived from the radial velocities based on the acoustic beam angle and direction. The timing of return echoes establishes the height at which the scattering occurred. Most sodar devices used for wind resource assessment measure the wind from 30 m up to about 200 m above ground in increments of 5–20 m. Figure 8-1 illustrates a sodar in operation, and Figure 8-2 shows two particular sodar models.

A typical sodar system is equipped with a series of speakers that function as transmitters and receivers, an onboard computer containing the operating and data processing software (including self-diagnostics), a power supply, and a combination data storage and communications package. Some sodars are trailer-mounted so they can be easily moved, and the trailers are partially enclosed for security and protection from the weather. The power supply must be sized to maintain continuous operation of the sodar and communications equipment. If the sodar is operated off-grid, a battery recharging system such as a diesel or gas generator, solar panels, or wind generator

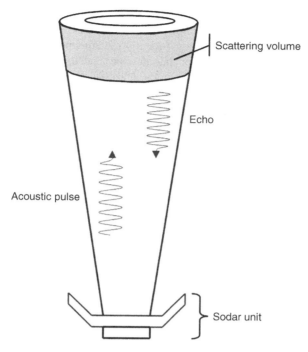

Figure 8-1. Illustration of sodar operation; the sodar unit emits an acoustic pulse and subsequently measures the frequency and time delay of the backscattered signal to determine the wind speed and height. *Source:* AWS Truepower.

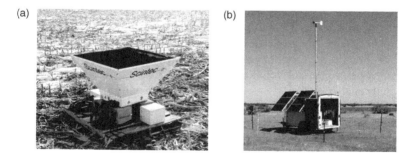

Figure 8-2. (a) Scintec SFAS sodar unit. (b) Atmospheric Research & Technology VT-1 sodar unit enclosed within a trailer. *Source:* AWS Truepower.

is needed. Sodar units (like lidar units) consume more power than most monitoring towers.

Sodar systems can require more complicated data-quality screening and analysis procedures than meteorological masts typically do. There are more parameters to check, differing system responses to atmospheric events (e.g. rain, snow, and other precipitation), and additional analyses to perform to obtain accurate results. Further analysis may also be required in complex flow conditions to obtain readings comparable to anemometer readings. It is consequently recommended that staff carrying out the analysis receive special training or that an experienced consultant be employed to carry out the data validation and preliminary analysis.

Air temperature and precipitation should be measured at the sodar site to facilitate data-quality screening and improve measurement accuracy. Air temperature is needed to accurately calculate the speed of sound, which in turn determines both the altitude assigned to returned echoes and the estimated tilt angle of a phased-array sodar's emitted acoustic beams. Precipitation can cause acoustic noise and scattering of sound back to the sodar. It can also invalidate the vertical velocity measurements. For these reasons, periods of measurable precipitation should be identified and will most likely have to be excluded from the data analysis.

8.2 LIDAR

Lidar operates by emitting a laser light signal (either as pulses or a continuous wave) that is partially scattered back in the direction of the emitter by aerosol particles suspended in the air. The light scattered from these particles is shifted in frequency, just as the sound frequency is shifted for a sodar system. This frequency shift is used to derive the radial wind speed along the laser path. Multiple laser measurements are taken at prescribed angles to resolve the 3D wind velocity components. The operational characteristics, number of measurement ranges, the depth of the observed layer, and even the shape of the measurement volume vary by lidar model type.

Two distinct types of lidar currently exist for wind resource assessment. *Profiling lidars* measure the wind along one dimension, usually vertically, producing a profile similar to that taken from a tower or sodar. These lidars typically measure wind speeds up to 200 m above the device. *Three-dimensional scanning lidars* have the capacity to direct the laser about two axes, which allows them to measure the wind throughout a hemispherical volume. Side-scanning devices are designed to obtain a three-dimensional grid of wind speeds over a large area, with some units having a range of several kilometers. While the side-scanning lidars have the potential to contribute greatly to wind resource assessment, this chapter focuses on the more widely tested and used profiling units.

A typical profiling lidar system is equipped with one or more laser emitters and receivers, an onboard computer containing the operating and data processing software (including self-diagnostics), heating and cooling systems, and a combination data storage and communications package. While most lidars come equipped to accept AC grid power and have onboard battery backup in case of a grid outage, a remote power supply must be acquired or custom-built for autonomous operation away from the grid. Like sodars, lidar units can be trailer-mounted and partially enclosed; however, most are sold by their manufacturers as stand-alone units. Figure 8-3 depicts two lidar models.

Figure 8-3. NRG/Leosphere's Windcube lidar (left), and Natural Power's ZephIR lidar unit. *Source:* AWS Truepower.

Lidars designed for wind energy applications came on the scene after sodars, and are considerably more expensive. Nevertheless, their popularity is growing, particularly in Europe, where most of the leading manufacturers are located. Lidars have benefited from testing campaigns that have helped to establish a reputation for accuracy. In addition, lidars are increasingly being considered for specialized applications, such as offshore wind resource assessment, replacements for nacelle anemometers on wind turbines, and deployment within and around existing wind farms for performance assessment. The use of lidar is expected to continue growing in the future as prices decrease and experience with and acceptance of the technology increase.

8.3 REMOTE SENSING CAMPAIGN DESIGN AND SITING

A successful remote sensing campaign, whether it uses sodar or lidar or both, requires considerable expertise in siting, system operations, and data analysis and interpretation.

Like meteorological masts, remote sensing systems should be placed at sites that are representative of wind conditions likely to be experienced by the wind turbines. The units should be installed level, and their orientation relative to true north should be determined and documented.

To prevent noise echoes that may harm data quality, sodars should be placed no closer to obstacles, such as meteorological masts, trees, or buildings, than the height of the obstacles. In many instances, and especially at sites with multiple surrounding obstacles (such as a clearing within a forest), a larger setback may be necessary. Rotating the sodar so that its acoustic beams are directed away from objects can sometimes reduce echoes. Nearby noise sources such as generators, air conditioners, and other emitters of high pitched tones should be avoided, if possible. Lastly, because the beeping or chirping can disturb people living nearby, the sodar should be sited at least 500 m from homes in open, flat terrain, and at least 350 m from homes in other terrain.

In theory, because a laser beam is more tightly focused than sound waves, lidar is less susceptible than sodar to interference (echoes) from nearby obstacles. This attribute may make it possible to use lidar in locations that would be troublesome for sodar and to obtain a better match to anemometer data by placing the system closer to the reference meteorological mast. Nevertheless, it is preferable to keep the device's "measurement cone" as unobstructed as possible, especially when it comes to moving objects such as tree branches, guy wires, and anemometers. While some lidar devices can tolerate the blockage of a significant portion of their field of view, this may reduce data recovery and increase the error margin in the observed wind speeds. Some lidar units can be rotated so that their beams are directed away from obstacles.

The horizontal wind speed derived from both sodars and lidars can be biased in complex terrain. This is because the radial readings are spaced increasingly far apart as the measurement height increases, and in complex terrain, the vertical component of the flow may not be homogeneous over the sampling volume. The bias can be as much as 5% in very complex terrain but is usually much less. This feature of remote sensing measurements, as well as possible corrections for it, is an area of active research (Section 8.6).

How much sodar or lidar data must be collected at a site and over what period depends on the wind conditions and the objectives of the study. Where the system is the sole source of wind measurement, at least a year of data collection (12 continuous months) is recommended, just as for monitoring towers. In the more common situation where there is a reference meteorological tower at the site, a much shorter period will usually suffice. Ideally, in this case, the data collected should span a representative range of speeds and directions. This can be accomplished in 4–6 weeks at most locations. The precision achieved can be estimated by comparing the speeds with simultaneous measurements at a nearby reference mast or by observing the number of samples in the important direction and speed bins. To further improve confidence in the observed profile, measurements should be taken at different times of year, especially at sites where strong seasonal variations in the wind resource, including speed, direction, and shear, are expected.

8.4 DATA COLLECTION AND PROCESSING

The primary outputs at each height for both sodar and lidar systems are the horizontal wind speed and direction, vertical wind speed, and their associated standard deviations. In addition, some indicator of signal quality, such as the signal-to-noise ratio (SNR), as well as the maximum height of reliable data, is usually provided. Understanding the definitions and thresholds for these parameters is useful for establishing appropriate data screening procedures and for identifying suspect data periods.

The recording interval should be compatible with that being used by other measurement systems with which the sodar or lidar readings will be compared (typically 10 min). Other averages, such as 60-min or daily means, can be calculated later, if desired. Clocks in the data recorders of all systems should be synchronized.

Sodar systems record a complete wind profile at each moment of time over a range of heights and at intervals determined by the software settings. The pulse repetition rate (or duty cycle) of the sodar is determined in part by the maximum measurement altitude. Increasing the altitude can reduce the number of valid data samples included in each recording interval. For example, for one common sodar type, setting the maximum altitude to 200 m typically results in about 15% fewer samples per 10-min recording interval than does setting the maximum altitude to 150 m. Since the SNR is related to sample size, this setting may influence data quality and data recovery, depending on the atmospheric conditions.

The elevations of the lidar range gates (height intervals over which data are recorded) can be programmed by the user, but the number of reporting levels is currently limited to between 5 and 10, depending on the model. Given the limited number of reporting elevations, the reporting heights should be chosen carefully. For example, two of the lidar range gates could be chosen to correspond with the top two anemometer heights on the reference wind monitoring tower to enable a direct comparison of the observed speeds. A third could be set at the expected turbine hub height, and the remainder spaced across the rotor plane.

8.5 COMPARISONS WITH CONVENTIONAL ANEMOMETRY

Since turbine power curves are currently defined with respect to wind speeds measured by IEC Class I anemometers, it is important that any source of bias in sodar or lidar readings with respect to such anemometers be understood and quantified.

Without the adjustments described below, sodar speeds can read 5–7% lower than anemometer speeds. Lidar speeds, too, can differ by up to 4–6% from cup anemometer readings at some sites. If the anemometers on the reference mast are not IEC Class I models, then it is equally important to understand their dynamic response characteristics and, if necessary, correct their readings to the IEC Class I standard.

The main factors responsible for biases between sodar, lidar, and cup anemometers are discussed below. Once these factors are accounted for, sodar and lidar speed measurements should usually fall within about 2% of concurrent measurements from a nearby anemometer at the same height.

8.5.1 Beam Tilt (Sodar)

The tilt angle of the acoustic beam emitted from a phased-array sodar varies slightly with the speed of sound through air, which is a function of temperature. Such variations can affect the accuracy of the derived speeds. Most sodar manufacturers address this issue by continually measuring the temperature at the sodar unit and computing the beam geometry. Temperature readings from a nearby mast can also be used in postprocessing the data. Failure to account for variations in temperature can result in biases of up to 2–3% between a sodar and a nearby anemometer.

8.5.2   Vector to Scalar Wind Speed Conversion (Sodar and Lidar)

Sodars and lidars typically record a vector-average horizontal wind speed at the end of each averaging period.[1] In turbulent conditions, the varying wind direction causes the vector average to be less than the scalar average (the usual output from anemometers). A correction to the vector average can be estimated if the standard deviation of the wind direction is known. Unfortunately, it is not possible to rely on the sodar's or lidar's own horizontal standard deviation readings to perform this correction because they, too, are based on vector averages. One option is to use the direction standard deviation recorded by an anemometer at a nearby mast. If this is not possible, then the standard deviation of the vertical speed measured by the sodar or lidar can be used instead. The vector-to-scalar bias correction is typically about 1–3%.

Some devices may apply a vector to scalar correction during data processing. This should be confirmed with the manufacturer.

8.5.3   Environmental Conditions (Lidar)

For all lidar devices, data recovery depends on the number and size of aerosol particles in the atmosphere. In especially clean air (e.g., high mountain air, and other environments after a rain storm), signal recovery at all monitoring heights is reduced, and measuring speeds at over 150 m height may not be possible. Some lidar devices are also sensitive to backscatter from clouds. Although corrective algorithms have been created for these conditions, this is an ongoing area of development. Finally, data collected during periods of precipitation should be excluded from certain analyses. While the effects of rain and snow on horizontal wind speed measurements may be small, the vertical wind measurements are almost always overwhelmed by the precipitation's downward motion and should be ignored during these periods.

8.5.4   Turbulence Intensity and Anemometer Overspeeding (Sodar and Lidar)

As noted in Chapter 4, cup anemometers tend to overestimate the mean wind speed in turbulence because they speed up in a gust more quickly than they slow down after the gust passes. This effect varies significantly by sensor model, and where the anemometers are not IEC Class I models, it can produce an apparent negative bias of as much as 1–3% in lidar or sodar readings compared to anemometer measurements. In this case, it is appropriate to adjust the anemometer data to avoid possibly overestimating power production by the turbine. This topic is addressed in Chapter 9.

---

[1] In a vector average, the east–west and north–south components of the speed are averaged separately and then converted to a magnitude. If the speed changes direction from, say, east to west within the averaging interval, the vector average can be much smaller in magnitude than the scalar average speed.

8.5.5 Flow Inclination and Complex Terrain (Sodar and Lidar)

Since sodars and lidars, unlike some cup anemometers, yield a true horizontal wind speed, it is necessary to determine if off-horizontal winds are contributing to apparent biases with respect to anemometer data. In most cases, the difference is less than 1%, but in extreme slopes, it may be as large as 3%. Since turbines respond only to the horizontal wind component, any adjustment should be applied to the anemometer data based on that sensor's response to inclined flow, as described in Chapter 9.

In some cases, particularly over narrow ridgelines and in other complex topography, flow inclination can be nonhomogeneous over the sodar or lidar measurement volume. This can cause discrepancies of 3–5% with nearby anemometer readings. The effect varies with the characteristics of the anemometers and remote sensing units, terrain complexity, land cover, and distance between measurements. This topic is an area of continuing investigation. Fully characterizing complex sites and reconciling anemometer and remotely sensed measurements may require bringing additional tools to bear, including high-resolution flow modeling and high-frequency 3D point and volume measurements.

8.5.6 Volume Averaging (Sodar and Lidar)

Both sodars and lidars measure the wind speed in a volume of air, in contrast to the "point" measurements of anemometers. Each layer measured by sodar (regardless of the height interval at which speeds are reported) actually represents an integral of information over a depth of 20 m or more. In layers where there is high wind shear, the volume averaging can cause the sodar to underestimate the mean speed at the measurement height by up to 3%.

For lidar, the depth of the volume measured can range from less than a meter to more than 50 m. The actual depth depends on the lidar type, and may be either variable or fixed over the entire profile. With greater volume depths, high shear can introduce a bias similar to that seen in sodar systems.

8.5.7 Distance from Reference Mast

Sometimes, sodars and lidars must be placed a considerable distance from a reference mast to minimize the mast's interference with the measurements, maximize data recovery, or meet other monitoring needs. In moderate and complex terrain and where there are significant variations in land cover, this can create apparent discrepancies between measurements simply because of the distance between the two locations. Assessing the significance of these discrepancies often requires expert judgment, sometimes aided by numerical wind flow modeling.

8.6 QUESTIONS FOR DISCUSSION

1. How do lidar and sodar work? Describe the basic operating principles of each.

2. In general, what advantages and disadvantages do lidar and sodar offer compared to a 60-m tilt-up tower? Discuss both positive and negative aspects for each type of remote sensing system.

3. Both lidar and sodar measure a "volume" of air. How is this different from an anemometer? What are the implications associated with this difference? Assume the wind speed increases across a range of heights from 90 to 110 m according to the power law (Chapter 3), with an exponent of 0.50. Assume further that the speed at 100 m is exactly 10 m/s. What is the average of the speeds at 90, 100, and 110 m? What percent error does this represent compared to the true 100-m speed?

4. What is "SNR," and how does it apply to ground-based remote sensing devices?

5. You are conducting a wind resource assessment campaign and want to determine how wind shear will affect your choice of turbine hub height. You decide to supplement your current assessment campaign, which uses two 60-m met towers, with a ground-based remote sensing device. The available monitoring location is near woods with 30-m-tall trees and a road with frequent traffic. Your budget allows for either a lidar or a sodar, and both are available from a supplier who can meet your schedule. (i) Which type of device would you choose? (ii) Why is the chosen device appropriate for the site? (iii) What siting considerations should be addressed during deployment?

6. Your candidate wind project site faces up a steep slope in the prevailing wind direction. What adjustments might be considered when comparing the wind data collected on a monitoring tower using cup anemometers with that collected with either lidar or sodar? Explain the item or parameter to be considered or adjusted for and if it applies to all remote sensing systems or just a particular type.

SUGGESTIONS FOR FURTHER READING

Deolia RA. Characterization of winds through the rotor plane using a phased array SODAR, Sandia National Laboratory, Report SAND2009-7895, Feb 2010. Available at http://prod.sandia.gov/techlib/access-control.cgi/2009/097895.pdf. (Accessed 2012).

Kelly ND, et al. Comparing pulsed doppler LIDAR with SODAR and direct measurements for wind assessment. National Renewable Energy Laboratory, Conference Paper CP-500-41792, 2005. Available at http://www.nrel.gov/wind/pdfs/41792.pdf. (Accessed 2012).

Moore K. Recommended practices for the use of sodar in wind energy resource assessment, Integrated Environmental Data, June 2010. Available at www.iedat.com/documents/SODARRecommendedPractices_IEARevised22June.doc. (Accessed 2012).

Weitkamp C, editor. Lidar: range-resolved optical remote sensing of the atmosphere, Springer series in optical sciences. 2005. p. 460.

Wharton S, Lundquist JK. Atmospheric stability impacts on power curves of tall wind turbines, Lawrence Livermore National Laboratory, Report LLNL-TR-424425, Feb 2010. Available at https://e-reports-ext.llnl.gov/pdf/387609.pdf. (Accessed 2012).

PART 2

DATA ANALYSIS AND RESOURCE ASSESSMENT

9

DATA VALIDATION

After the wind resource measurements are collected and transferred to an office computing environment, the next step is to quality control (QC) and validate the data. The purpose of this process is to ensure that only valid data are used in subsequent analyses and that the data are as accurate as possible. Problems overlooked at this stage can result in significant errors in the wind project's estimated energy production.

While different analysts use different terminology, QC generally refers to the initial screening of data for obvious problems such as logger and sensor failures and data transmission failures. This should be done as soon as possible after the data are transferred from the logger to ensure that instrument problems are discovered, and fixed, promptly.

Data validation, a more involved process, is done less frequently (typically monthly or quarterly). Validation means the inspection of data for completeness and reasonableness and the detection and flagging of bad (invalid or suspect) values in the data record. A number of methods, which are described in detail in this chapter, can be used. It should be noted, however, that no data validation procedure is likely to catch every bad record, and moreover, good data may sometimes be wrongly rejected. Data validation is like any statistical decision process subject to both type I (false positive)

Wind Resource Assessment: A Practical Guide to Developing a Wind Project, First Edition.
Michael Brower et al.
© 2012 John Wiley & Sons, Inc. Published 2012 by John Wiley & Sons, Inc.

and type II (false negative) errors. Sometimes methods designed to minimize one type of error result in an excessive number of the other type. A good data validation procedure seeks to minimize both types of error.

In this chapter, techniques appropriate for both QC and data validation are covered under the validation process.

The validation of remotely sensed data is a more specialized topic, which is outside the scope of this book.

9.1   DATA CONVERSION

Depending on the data logger manufacturer and model, the data may first need to be converted from the logger's raw binary format to a text file, spreadsheet, database, or some other usable format. Manufacturers of the most widely used data loggers (e.g., Campbell Scientific, NRG Systems, Second Wind) provide software to do this, which is either part of the logger software or runs on a separate computer.

In performing this conversion, the analyst must make sure that settings such as the wind vane deadband, anemometer transfer function, and time zone are correctly entered in the conversion software. This may seem like a trivial requirement, but surprisingly many mistakes occur at this stage. For example, it is not uncommon for boom orientations and magnetic declinations to be entered incorrectly in the site documentation or for sensor channel numbers to be switched. These and other common mistakes, if not caught at the outset, can lead to significant errors in characterizing the site's wind resource.

For this reason, as a general rule, the analyst should seek independent confirmation of key information whenever possible. For example, photographs can help confirm reported sensor heights and boom lengths and orientations, and scatter plots of the ratios by direction of speeds from paired anemometers can help verify anemometer boom orientations and designations. If no detailed site documentation is available—or if the documentation was provided by another party—a visit to the site to obtain or confirm the required information might be warranted.

Calibrated anemometers should be accompanied by a certificate from the agency that performed the calibration test. The analyst should check this certificate to confirm the sensor transfer function and to verify that the sensor test was normal. There is currently some debate within the wind industry about whether, for calibrated anemometers, the measured transfer function or an average "consensus" function based on numerous tests of different anemometers of the same model should be used when converting raw data. Either method is generally acceptable, although there is evidence that for NRG #40 and Second Wind C3 cup anemometers, in particular, the consensus transfer function yields results that tend to match IEC Class I anemometers employed for power curve testing more closely than the measured functions do (1, 2).

As a matter of good data handling practice, both the raw and converted data should be preserved in permanent archives. All subsequent data validation and analyses should be performed on copies of the converted data files. Different file name extensions should be used to avoid confusion. For example, raw data can be given the extension *raw*, while verified data can be given the extension *ver*.

9.2   DATA VALIDATION

In these days of powerful personal computers, most data validation is done with automated tools; however, a manual review is still highly recommended. Validation software can be obtained from some data logger vendors, and commercial software is also available. Firms that do a lot of data validation often create their own automated methods using spreadsheets or custom software written in languages such as Fortran, Visual Basic, C++, or R.

Whatever method is used, data validation usually proceeds in two phases: automated screening and in-depth review. The automated screening uses a series of algorithms to flag suspect data records. Suspect records contain values that fall outside the normal range based on either prior knowledge or information from other sensors on the same tower. The algorithms commonly include relational tests, range tests, and trend tests.

The second phase, sometimes called *verification*, involves a case-by-case decision about what to do with the suspect values—retain them as valid or reject them as invalid. This is where judgment by an experienced person familiar with the monitoring equipment and local meteorology is most helpful. Information that is not part of the automated screening, such as regional weather data, may also be brought into play.

As an example of how this process can unfold, the automated screening might flag a brief series of 10-min wind speeds as questionable because they are much higher than the speeds immediately before and after. Was this spike real, or was it caused by a glitch in the logger electronics, such as might be caused by a loose connection? During the review phase, the reviewer might check other sensors on the same mast and observe the same spike; this would suggest that it is not a problem with a single sensor or logger channel. Then he or she might look at regional weather records and find that there was thunderstorm activity in the area at the time. The conclusion is that the spike was most likely caused by a passing thunderstorm and should not be excluded from the data analysis.

Another example is presented in Figure 9-1. After a period of apparently normal operation, the 10-min average speed readings from an anemometer dropped to the off-set value (indicating no detectable wind), while the standard deviation dropped to zero. Later, both appeared to return to their normal behavior. The reviewer checks the temperature and finds it hovered near freezing before the event and rose above freezing at the end. Furthermore, the direction standard deviation (not shown) fell to zero shortly before the speed standard deviation did and resumed normal behavior at about the same time. The conclusion is that this was a likely icing event and should be excluded.

In such a two-phase validation approach, it is reasonable for the automated screening to be somewhat overly sensitive, meaning it produces a greater number of false positives (data flagged as bad, although they are actually good) than false negatives (data that are cleared as good but are actually bad). One reason for this bias toward overdetection is that there will be an opportunity to reexamine bad data records in

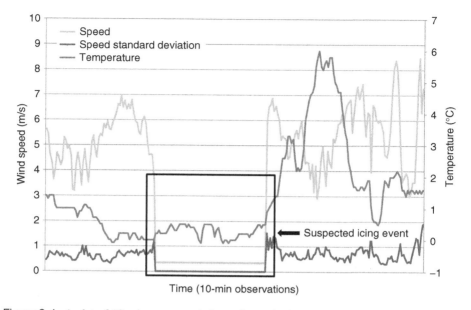

<u>Figure 9-1.</u> A plot of 10-min average wind speed, standard deviation of speed, and temperature showing a suspected icing event. Time series plots such as these can be a valuable tool for verification of suspected bad data. *Source*: AWS Truepower.

the review phase, whereas good records usually receive no further scrutiny. Another reason is that failing to reject even a small number of bad values can significantly bias a wind resource analysis, whereas excluding a moderate amount of good data rarely has such an impact. However, care must be taken in designing the automated screening so as not to overwhelm the review phase with an excessive number of false positives. Finding the right balance takes trial and error.

9.2.1 Validation Routines

Validation routines are designed to screen each measured parameter and flag suspect values for review. They can be grouped into two main categories: general system checks and measured parameter checks.

General System Checks. Two simple tests described below assess the completeness of the collected data:

Data Records. The number of data fields must equal the expected number of measured parameters for each record.

Time Sequence. The time and date stamp of each data record are examined to see if there are any missing or out-of-sequence data.

Table 9-1. Examples of range test criteria

Parameter[a]	Validation criteria
Wind Speed: Horizontal	
Average	Offset < Avg. < 30 m/s
Standard deviation	0 < Std. Dev. < 3 m/s
Maximum gust	Offset < Max. < 35 m/s
Wind Direction	
Average	0° < Avg. < 360°
Standard deviation	3° < Std. Dev. < 75°
Temperature	Varies seasonally
Typical range	−35° < Avg. < 35°C
Wind Speed: Vertical	Varies with terrain
Average (S or C)	Offset < Avg. < ± (2 or 4) m/s
Standard deviation (S or C)	Offset < Std. Dev. < ± (1 or 2) m/s
Maximum gust (S or C)	Offset < Max. < ± (3 or 6) m/s
Barometric Pressure	Sea level shown
Average	94 kPa < Avg. < 106 kPa
Differential Temperature	
Average difference	>1.0°C (daytime)
Average difference	<1.0°C (overnight)

Abbreviations: S, simple terrain; C, complex terrain. *Source:* AWS Truepower.
[a] All monitoring levels except where noted.

Measured Parameter Checks. Three measurement parameter checks are commonly performed: range tests, relational tests, and trend tests. These tests are applied in sequence, and data must pass all three to be deemed valid.

RANGE TESTS. In range tests, the measured data are compared to allowable upper and lower limiting values.[1] This is the simplest and most common type of test. Table 9-1 presents examples of range test criteria. A reasonable range for 10-min average wind speeds is from a minimum of the anemometer offset to a maximum of 30 m/s. Any values that fall below the anemometer offset should be flagged as either missing or invalid; speeds above 30 m/s are possible but should be verified. The limits of each range test should be set so they span nearly the full range of plausible values for the site. In addition, the limits should be adjusted seasonally, where applicable. For instance, the limits for air temperature and solar radiation should be lower in winter than in summer.

RELATIONAL TESTS. These tests rely on relationships between various measured parameters. For example, wind speeds recorded at the same height should be similar

---

[1] Note that a variety of voltage measurement systems exist, which are intended to measure different systems (communications device, internal battery voltage, external power source measurement). Each system has different operating ranges, and care should be exercised when creating range and relational tests for these devices.

(except when one anemometer is in the tower shadow); wind shears between heights should fall within reasonable bounds (which may vary diurnally and seasonally). Table 9-2 gives examples of several relational test criteria for 10-min data. These tests should ensure that physically improbable situations (such as a significantly higher speed at 25 m compared to 40 m) are scrutinized. Comparisons between paired sensors at the same height are especially valuable.

Scatter plots of the speed ratio between a pair of sensors at the same height as a function of speed can be helpful for detecting problems. Figure 9-2 illustrates two cases, the first exhibiting a normal degree of scatter and the second an abnormal degree caused most likely by degradation or damage in one or both of the anemometers.

Table 9-2. Examples of relational test criteria

Parameter[a]	Validation criteria
Wind Speed	
Max. gust vs average	Max. gust $\leq 2.5 \times$ Avg.
60/40 m Average difference	≤ 3 m/s
60/40 m Daily Max. difference	≤ 5 m/s
60/25 m Average difference	≤ 5 m/s
60/25 m Daily Max. difference	≤ 8 m/s
Wind Speed: Same Height	
Average difference	≤ 0.5 m/s
Maximum difference	≤ 3.0 m/s
Wind Direction	
60/25 m Average difference	$\leq 20°$
Wind Shear	Varies with terrain
60/25 m Average	$-0.05 < \alpha^b < 0.45$

[a] All monitoring levels except where noted.
Source: AWS Truepower.
[b] Wind shear exponent.

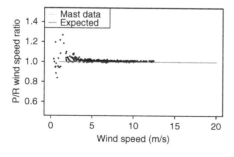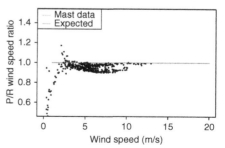

Figure 9-2. Plots of speed ratios as a function of wind speed for pairs of anemometers at the same height. (a) A generally tight relationship with a normal degree of scatter at low speeds. (b) This chart suggests that at least one of the two sensors is not performing to specification. *Source*: AWS Truepower.

Table 9-3. Examples of trend test criteria

Sample parameter[a]	Validation criteria
Wind Speed Average	All sensor types
1-h Change	<5.0 m/s
Temperature Average	
1-h Change	≤5°C
3-h Change	≤1kPa
Differential Temperature	
3-h Change	Transitions twice above
	or below freezing

[a] All monitoring levels except where noted.
Source: AWS Truepower.

TREND TESTS. These checks are based on the rate of change in a value over time. Table 9-3 lists sample trend test criteria. The thresholds actually used should be adjusted as necessary to suit the site conditions. Note that wind direction trends are not considered because direction can change abruptly during severe weather or frontal passage events, among other conditions.

The examples of validation criteria in Tables 9-1–9-3 are neither exhaustive nor do they necessarily apply to all sites. With experience, the analyst will learn which criteria are most useful in particular conditions.

In addition to these standard tests, two situations usually receive special flags: tower shadow and icing.

TOWER SHADOW. Tower shadow is identified when two anemometers fail a relational test and the wind is from a direction in which one of them is downwind of the mast. The angular width of the zone of tower shadow depends on the geometry of the mast but is typically about 30° on either side of a line directly through the tower. For example, if the boom points due east from the mast, wind directions from 240° to 300° would be flagged. The shadowed region may be different for a lattice tower because the boom is typically offset from the center of the tower. Before applying such a decision rule, it is a good idea to verify the direction of peak shadow and the width of the shaded zone by plotting the ratio of speeds between two anemometers at the same height as a function of wind direction (Fig. 9-3). Such plots can also reveal unexpected influences by the tower and equipment mounted on it.

ICING. Icing events are usually flagged when the standard deviation recorded by the direction vanes is zero or near zero, and the temperature is near or below freezing. This is a conservative approach since direction vanes tend to freeze before anemometers do. During periods of detectable icing, it is unwise to rely on data from unheated anemometers even if the speeds are above the offset, since they could be slowed by moderate ice accumulation.

9.2.2 Treatment of Suspect Data

After the raw data are subjected to the automated validation checks, a reviewer should decide what to do about the suspect data records. Some suspect values may represent real (albeit unusual) weather occurrences, which should not be excluded from the resource assessment, while others may reflect sensor or logger problems and should be eliminated.

Here are some guidelines for handling suspect data:

- Check to see whether data from different sensors on the same mast confirm the suspect reading. If a transient feature such as a large jump in wind speed is noted at one anemometer, is a similar jump seen at other anemometers? If only one sensor shows the feature, it is more likely that the data for that sensor are invalid.

- Use data from a variety of sources to verify weather conditions. If icing is suspected, is this supported by the observed temperature? If large changes in wind or temperature are seen in the record, do local weather stations indicate a passing weather front that might explain the pattern?

- Examine relationships between sensors over time. Very often, sensor degradation happens so slowly that it goes unnoticed if the data are only examined in periods of, say, 2 weeks or a month at a time. By examining the relationships over several months or longer, the degradation becomes obvious. Other problems, such as icing, take a limited time to develop and disappear and, moreover, may not affect sensors at different heights to the same degree. Periods around flagged icing episodes should be scrutinized carefully, however, to be sure the times of onset and conclusion have been accurately identified. This is because anemometers sometimes experience slowdown before the thresholds signaling an icing event are crossed.

- Assign invalid data a code indicating the suspected reason. Table 9-4 gives some examples of validation codes. An examination of operation and maintenance logs, site temperature data, and data transmission logs may help determine the appropriate code.

- Maintain a complete record of all data validation actions for each monitoring site in a log file.

Table 9-4. Examples of validation codes

Code	Rejection criteria
−990	Unknown event
−991	Icing or wet snow event
−992	Static voltage discharge
−993	Wind shading from tower
−995	Wind vane deadband
−996	Operator error
−997	Equipment malfunction
−998	Equipment service
−999	Missing data (no value possible)

9.3 POST-VALIDATION ADJUSTMENTS

Good sensors mounted correctly should provide accurate measurements of wind speed, direction, and other meteorological parameters most of the time. However, there are several factors that often need to be considered to accurately estimate the true free-stream speed. This section addresses three types of adjustment: tower effects, turbulence, and inclined flow. Some adjustments apply to only certain types of anemometers.

9.3.1 Tower Effects

Even outside the zone of direct tower shadow, the presence of the tower can increase or decrease the observed wind speed compared to the true free-stream speed. The effect depends on direction, the sensor's distance from the tower, and the tower width and type. For a "goal post" configuration above the tower, the effect may be negligible. Directly upwind, a tower impedes the wind, reducing the speed; over certain angles on either side of the tower, the tower causes the wind flow to accelerate, producing an increase in the observed speed (refer to Fig. 5-7). These effects are often readily apparent in a scatter plot of speed ratios by direction for a pair of sensors at the same height (Fig. 9-3).

Depending on the boom length and tower geometry, secondary tower effects such as these can be up to several percent, a significant impact for resource assessment, especially if the wind comes often from a narrow range of directions. For example, it was once quite common to place anemometer booms 180° apart and perpendicular to the prevailing wind direction. For a tubular tower, this configuration tends to result in an overestimate of the free-stream mean speed at both sensors.

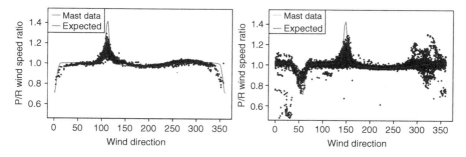

Figure 9-3. Plots of speed ratios as a function of wind direction for pairs of anemometers at the same height. The red lines are the result of a computational fluid dynamical model of the tower effects. The prominent dips and spikes in both charts represent the effect of tower shadow. Note that the implied boom direction is 180° opposite the wind direction. (a) A normal degree of scatter and secondary tower influences outside the shadow directions. (b) Relatively large secondary influences, which may be due to equipment or obstructions on the tower. *Source*: AWS Truepower.

By correcting for these tower influences, a more accurate free-stream speed reading can be obtained for an individual sensor. Currently, however, there are no commercial tools available for doing this, so custom tools must be developed by the resource analyst using information available in the literature (3). As an alternative, averaging valid data from two sensors at the same height and oriented at the recommended angular distance apart (depending on tower type) usually mitigates and can virtually eliminate tower effects in the combined data record. Data averaging is discussed in more detail in Section 9.4.2.

9.3.2 Turbulence

Cup anemometers are known to overestimate the wind speed in turbulent flow conditions because of the tendency of the anemometers to respond more quickly to abrupt increases in speed than to rapid slowdowns. The magnitude of the overspeeding depends on the sensor type and degree of turbulence.

Research has shown that anemometers with a relatively large distance constant, such as the NRG #40 and Second Wind C3, record greater wind speeds in turbulent conditions than do IEC Class I anemometers used for turbine power performance testing and certification. When turbulence is low, the opposite tendency can occur. By applying a sensor-specific adjustment to account for these tendencies, a more accurate energy production estimate can be obtained (4).

In contrast to cup anemometers, prop-vane anemometers tend to underestimate the wind speed as a result of turbulence. This is because the wind direction changes so quickly that the vane cannot keep the propeller aligned perfectly into the wind. A propeller anemometer only measures the component of the wind speed that is parallel to its rotation axis; the observed speed is reduced by a factor equal to the cosine of the angle of deviation. Since greater turbulence produces larger direction shifts, the magnitude of prop-vane underspeeding typically increases with increasing turbulence intensity (5).

Sonic anemometers, lacking moving parts, are insensitive to turbulence. To bring their measurements in line with those of Class I sensors, however, an adjustment for turbulence may nonetheless be called for, although it is usually small.

9.3.3 Inclined Flow

Horizontal axis wind turbines generate power from the component of the wind that is perpendicular to the turbine rotor's plane of rotation. To support an accurate energy production estimate, anemometers should ideally measure only that component. However, cup anemometers, in particular, are sensitive to varying degrees of off-horizontal winds depending on the geometry of the cups and instrument. Research has documented the impact of flow angle on wind speeds recorded by cup anemometers of various types (6), but making use of this information requires knowledge of the flow angle at the tower. This can be obtained from a sodar, a lidar, or a vertical anemometer mounted on the mast. Without such a direct measurement, the flow angle can be estimated from the terrain slope and from wind flow modeling. Inclined flow can also occur in flatter terrain for brief periods under low wind and strong surface heating, but this effect is usually small and thus requires no correction.

9.4 DATA SUBSTITUTION AND AVERAGING

Up to this point, the data validation process has sought to keep valid data from each sensor intact and separate from the data from other sensors on the same tower. In this section, two methods of combining the data from different sensors are discussed: substitution and averaging. Data substitution aims to create the longest possible data record by filling gaps in one sensor's record with data from one or more other sensors; data averaging seeks to reduce the uncertainty in the observed speeds by combining measurements from two different anemometers at the same height.

9.4.1 Data Substitution

Since a key objective of the wind resource monitoring program is to develop a time series of wind data covering as long a period as possible, it is desirable to fill any gaps in the record with valid data from other sensors when available. Data substitution is virtually a requirement for anemometers at the top mast height, as well as for the top direction vanes, as they are the most important for assessing the site's wind resource. Whether data substitution is performed for lower level anemometers or temperature and pressure sensors is largely a matter of preference (note that for reasons discussed in Chapter 10, substituted data should not be used for estimating the wind shear).

For anemometers, the substituted data ideally should come from an instrument at the same height, although in rare instances—such as when both anemometers at the top height have malfunctioned for an extended period—data from an anemometer at a different height may be used. In any case, before the substitution is carried out, a relationship (such as a linear regression forced through the origin or a simple ratio) between the two anemometers should be established from concurrent, valid data. The analyst should verify that the relationship between them is tight and linear, as otherwise the results will be unreliable. This "field calibration" is especially important when there is a significant, persistent bias between the anemometer readings, which can happen with anemometers of different types (such as heated and unheated) and with anemometers at different heights.

It is generally straightforward to fill gaps in the directional data record using valid data from another vane. The analyst should merely check to make sure that there is no significant, persistent bias between the two vanes' directional readings during periods when both produce valid data. Such a bias could indicate a discrepancy in the boom orientations or vane deadbands and should be investigated and corrected, if possible. Note that large transient deviations in direction can occasionally arise under light, variable winds; when the wind is strong, however, the directions recorded at heights within 20–30 m of each other should be nearly equal (within $5°$).

9.4.2   Data Averaging

When anemometers are mounted in pairs at each height on the tower, the question arises, should both sets of measurements be used in characterizing the wind resource, and if so, how should they be combined?

A popular approach is to designate one of each pair of anemometers as the primary sensor and the other as the secondary sensor. The primary anemometer's data are used exclusively for the analysis except when they have been flagged as suspect or invalid. In those periods, the flagged data are replaced by valid data from the secondary sensor in the manner described in the previous section. (If no valid redundant data are available, a gap is left in the data record.)

The underlying assumption of this approach is that the primary sensor is the more accurate of the two. That may be a reasonable assumption in some cases—for example, when the secondary sensor is a heated cup anemometer (heated cup anemometers being generally less accurate than unheated cup anemometers, except, of course, in freezing conditions); when the primary sensor is of superior quality; or when the secondary sensor is in the tower shadow far more often than the primary sensor.

Very often, however, there is no reason to expect either sensor to be more accurate than the other most of the time, so the choice of primary sensor is arbitrary. The preferred method is then to average the data from the two anemometers. Assuming the measurement errors of the two sensors are uncorrelated and of roughly the same magnitude, this method reduces the uncertainty in the observed speed by a factor of $\sqrt{2}$, or 1.414, compared to relying on the data from one sensor alone. Averaging can also help mitigate secondary tower effects.

Averaging can be used only when the data from both sensors are valid; whenever one is shadowed or experiences some other problem, only the other's data should be counted. In those periods, the uncertainty reverts back to the uncertainty of the solitary sensor. (Uncertainty in resource assessment is covered in Chapter 15.)

9.5   QUESTIONS FOR DISCUSSION

1. Why is it sensible for an automated screening routine to produce more false positives than false negatives?

2. Why is boom orientation accuracy important? Name two possible impacts on the accuracy of energy production estimates if the assumed boom orientation is wrong. What are some ways one can independently verify the anemometer boom orientations given in the site documentation?

3. Describe a scenario in which a meteorological mast could artificially increase the average wind speed recorded by an anemometer. In what scenario might it be decreased? (Ignore tower shadow.)

4. Suppose the wind speed values for a given period of time equal the offset in the conversion equation and the standard deviation is zero. What might be causing this phenomenon, and what are some ways to determine the cause?

5. Why do you think wind vanes tend to freeze before anemometers?

6. Explain the differences between range tests, relational tests, and trend tests. List a few examples of each.

7. Explain the difference between QC and validation of meteorological data. Why are they both important?

8. Define the distance constant for an anemometer. How can the distance constant impact wind speed measurements?

9. Suppose you plot the ratio of wind speeds by direction recorded by two anemometers from the same level, and the resulting angles where tower shadow is apparent do not match expectations from the site information. What could be the problem? What are some methods to determine the source of the problem?

10. Describe in physical terms how atmospheric turbulence can impact the observed mean wind speed. Why do some anemometers measure greater wind speeds in high turbulence environments and lower wind speeds in low turbulence environments than an IEC Class I sensor? How about prop-vane anemometers?

11. Why is data averaging generally preferred over data substitution? Describe a scenario where data substitution would be required.

REFERENCES

1. Hale E. Memorandum: NRG #40 transfer function validation and recommendation, AWS Truewind, 8 Jan 2010.
2. Young M, Babij N. Field measurements comparing the Riso P2546A anemometer to the NRG #40 anemometer. Global Energy Concepts 2007.
3. IEC 61400-12-1. ed. 1.0. Geneva, Switzerland: IEC. Copyright © 2005.
4. Filippelli MV, et al. Adjustment of anemometer readings for energy production estimates. In: Proceedings of Windpower 2008, Houston, Texas, USA, June 2008.
5. Tangler J, et al. Measured and predicted rotor performance for the SERI advanced wind turbine blades. In: Proceedings of Windpower 1991, Palm Springs, California, USA, Feb 1992.
6. Papadopoulos KH, et al. Effects of turbulence and flow inclination on the performance of cup anemometers in the field. Boundary-Layer Meteorol 2001;101(1):77–107.

SUGGESTIONS FOR FURTHER READING

DeGaetano A. A quality control routine for hourly wind observations. J Atmos Oceanic Technol 1997;14:308–317.

Fiebrich C, Morgan C, McCombs A. Quality assurance procedures for mesoscale meteorological data. J Atmos Oceanic Technol 2010;27:1565–1582.

Makkonen L, Lehtonen P, Helle L. Anemometry in icing conditions. J Atmos Oceanic Technol 2001;18:1457–1469.

10

CHARACTERIZING THE OBSERVED WIND RESOURCE

Once the data validation is complete, the data can be analyzed to produce a variety of wind resource statistics and informative reports. This type of analysis provides a useful summary of the wind resource observed over the course of the monitoring program. Software to do this is available from several vendors, including some data logger manufacturers. Customized reports can also be created with spreadsheet and database software.

10.1 SUMMARIZING THE OBSERVED WIND RESOURCE

Table 10-1 presents a list of the summary statistics that are commonly provided in wind resource reports. These statistics, or some subset of them, may be generated on a periodic basis, such as monthly or quarterly, as well as annually and at the end of the monitoring program. A sample monthly report is provided in Figure 10-1.

The following sections describe how the various parameters are derived.

Wind Resource Assessment: A Practical Guide to Developing a Wind Project, First Edition.
Michael Brower et al.
© 2012 John Wiley & Sons, Inc. Published 2012 by John Wiley & Sons, Inc.

Table 10-1. Sample wind resource report statistics

Report products	Units
Data recovery (DR) fraction	%
Mean and annualized mean wind speed	m/s
Mean wind power density (WPD)	W/m^2
Wind shear	Nondimensional exponent
Turbulence intensity (TI)	%
Mean air temperature	°C
Mean air density	kg/m^3
Speed frequency distribution	Graph
Weibull A and k parameters	m/s (A), nondimensional (k)
Wind rose	Graph
Daily and hourly speed distributions	Graph

10.1.1   Data Recovery

The data recovery (DR) is defined as the number of valid data records (N_{valid}) divided by the total possible number of records (N) for the reporting period. It is usually expressed as a percentage. The equation is as follows:

$$DR = 100 \times \frac{N_{valid}}{N} (\%) \tag{10.1}$$

For example, the total possible number of 10-min records in December is 4464. If 264 records were deemed invalid, the number of valid data records collected would be 4200 (4464−264). The DR for this example would be

$$DR = 100 \times \left(\frac{4200}{4464}\right) = 94.1\% \tag{10.2}$$

The DR should be determined for each sensor for all levels at each site.

10.1.2   Mean and Annualized Mean Wind Speeds

The mean wind speed is simply the average of the valid speed values for the period in question:

$$\overline{v} = \frac{1}{N_{valid}} \sum_{i=1}^{N_{valid}} v_i \tag{10.3}$$

However, the mean wind speed can sometimes be a misleading indicator of the wind resource. If the data span a period much shorter than a full year, the mean will not reflect the full seasonal cycle of wind variations. Even if the data span a full year, there may be large gaps in the record that can bias the mean in favor of months with

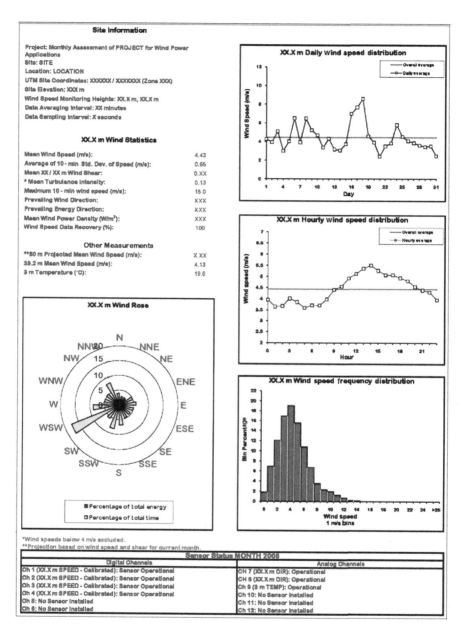

Figure 10-1. Sample summary of monthly wind statistics. *Source*: AWS Truepower.

more complete data coverage. And if the data cover more than 1 year but not an integer number of years, some calendar months may occur more often than others, also possibly resulting in a bias in the estimated mean speed.

As an example, suppose the data record spans 14 months, with January and February appearing twice. If the winter months are typically the windiest, a simple average of the wind speeds for the entire period will probably overestimate the true annual average speed. Or suppose the period of record spans exactly 12 months, but half the data in the winter months are lost because of icing. In that case, it is likely the calculated mean speed will understate the true annual average.

The *annualized mean wind speed* attempts to correct for these problems. Note that this is not the long-term historical mean speed, rather it is a seasonal correction for the observed period of data. (The long-term adjustment process is explained in Chapter 12.) The annualized mean can be estimated in a variety of ways but is usually found by calculating, first, the mean for each calendar month in the record, and second, the mean of the monthly means weighted by the number of days in each month. In equation form,

$$\overline{v}_{annual} = \frac{1}{365.25} \sum_{m=1}^{12} D_m \overline{v_m} = \frac{1}{365.25} \sum_{m=1}^{12} D_m \left(\frac{1}{N_m} \sum_{i=1}^{N_m} v_{im} \right) \qquad (10.4)$$

The outer sum is over the 12 calendar months, with D_m being the average number of days in month m (28.25 for February, counting leap years). The inner sum is over those speeds that fall within a particular calendar month. The calculation is illustrated in Table 10-2. Here, the data record spans 17 months, from January 2008 to May 2009, with January through May repeated. The straight average of the speeds (taking into

Table 10-2. Sample monthly data record for a station, illustrating the difference between period of record and annualized mean speeds

Month	No. of days in month	2008	2009	Average
January	31	8.94	8.68	8.81
February	28.25	8.35	7.37	7.86
March	31	7.63	8.13	7.88
April	30	6.79	7.00	6.90
May	31	6.56	6.85	6.71
June	30	6.58	—	6.58
July	31	5.81	—	5.81
August	31	6.25	—	6.25
September	30	7.50	—	7.50
October	31	7.85	—	7.85
November	30	8.26	—	8.26
December	31	8.36	—	8.36
Annualized average				7.39

account the DR in each month) is 7.49 m/s. However, the annualized average is only 7.39 m/s because the repeated months are windier, on average, than the other months.

Naturally, this method only works if the data record spans at least 12 months; although if it is only 1 or 2 months short of 12, an approximate annualized mean can sometimes be obtained by assuming that the missing months are similar to the months immediately before and after. The method can be applied to other parameters such as shear in a similar way.

10.1.3 Wind Shear

The wind shear (the rate of change in horizontal wind speed with height) is typically expressed as a dimensionless power law exponent known as *alpha* (α). The power law equation relates the wind speeds at two different heights in the following manner:

$$\frac{v_2}{v_1} = \left(\frac{h_2}{h_1}\right)^{\alpha} \tag{10.5}$$

where
$\quad v_2 =$ the wind speed at height h_2;
$\quad v_1 =$ the wind speed at height h_1.

This equation can be inverted to define α in terms of the measured mean speeds and heights:

$$\overline{\alpha} = \frac{\log\left(\frac{\overline{v_2}}{\overline{v_1}}\right)}{\log\left(\frac{h_2}{h_1}\right)} \tag{10.6}$$

Figure 10-2 depicts wind speed profiles for a range of exponents assuming a speed of 8.5 m/s at 120 m height.

Taking the average of the speeds before calculating the shear, as in the equation above, conveniently yields a time-averaged shear exponent, which is of most interest at this stage of the analysis. Time-averaged exponents can range from less than 0.10 to more than 0.40, depending on land cover, topography, time of day, and other factors. For short periods, and especially in light, unsteady winds, shear exponents can extend well beyond this range. Typical mean shear values are shown in Table 10-3 for a range of site conditions (this table is reproduced from Chapter 3). All other things being equal, taller vegetation and obstacles lead to greater shear. Complex terrain also usually produces greater shear, except on exposed ridges and mountain tops where topographically driven acceleration can reduce shear. Sites in tropical climates tend to have lower shear than similar sites in temperate climates because the atmosphere is less often thermally stable. (The effect of thermal stability is discussed in the next chapter.)

The calculated shear is sensitive to small errors in the relative speed between the two heights, and this sensitivity increases as the ratio of the two heights, h_1 and h_2,

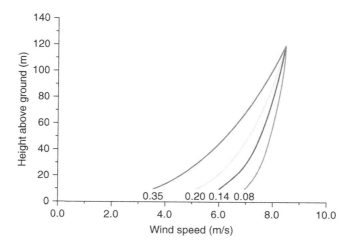

<u>Figure 10-2.</u> Theoretical profiles of wind speed with height for a range of values of the exponent α (0.08, 0.14, 0.20, and 0.35). All curves assume a speed of 8.5 m/s at 120 m height. *Source:* AWS Truepower.

Table 10-3. Typical shear exponents for different site conditions. These estimates may not be valid for specific sites; measurement is required

Terrain type	Land cover	Approximate range of annual mean wind shear exponent
Flat or rolling	Low to moderate vegetation	0.12–0.25
Flat or rolling	Patchy woods or forest	0.25–0.40
Complex, valley (sheltered)	Varied	0.25–0.60
Complex, valley (gap or thermal flow)	Varied	0.10–0.20
Complex, ridgeline	Low to moderate vegetation	0.15–0.25
Complex, ridgeline	Forest	0.20–0.35
Offshore, temperate	Water	0.10–0.15
Offshore, tropical	Water	0.07–0.10

Source: AWS Truepower.

decreases. This is evident from the following equation for the error in the shear exponent:

$$\Delta\alpha \cong \frac{\log(1+\varepsilon)}{\log\left(\dfrac{h_2}{h_1}\right)} \tag{10.7}$$

where ε is the error in the ratio of speeds measured by the top and bottom anemometers. For example, for anemometers at 40 and 60 m height (a height ratio of 1.5), an error of 1.5% in the speed ratio—a reasonable deviation under field conditions—results in an error of 0.037 in the shear exponent. This, in turn, produces an error of 1.1% in the predicted speed at 80 m. For heights of 50 and 60 m (height ratio 1.2), the same relative speed error produces an error of 0.082 in the exponent and 2.4% in the speed at 80 m.

Given the sensitivity of the calculated shear to small speed errors, three rules should be followed to produce a reasonably accurate shear estimate: First, the speed ratio should only be calculated using concurrent, valid speed records at both heights. This avoids errors caused by mixing data from different periods or with different rates of DR. Second, the two heights in the shear calculation should be separated by a ratio of at least 1.5 (e.g., 33 and 50 m or 40 and 60 m). Third, the speed data should originate from anemometers mounted on horizontal booms of the same length and with the same directional orientation relative to the tower, so that the effects of the tower on the speed observations will be similar. One implication of this last rule is that, in general, data that have been substituted from other sensors should not be used in shear calculations. Instead, only data originally collected from two identically oriented anemometers are appropriate for this purpose.

Just one average shear value for each pair of heights is usually provided in wind resource reports. This shear is calculated as noted above, by averaging the concurrent speeds from each anemometer, then taking the ratio and calculating the exponent. Some analysts choose to exclude speeds below 3 or 4 m/s in this calculation, as shear tends to be more variable in light winds, and low speeds do not contribute significantly to energy production. In the following chapter, the use of instantaneous or binned shear exponents to extrapolate a time series of wind speed data to hub height, along with possible adjustments to the shear above the top anemometer height, is discussed.

10.1.4 Turbulence Intensity

Wind turbulence, defined as rapid fluctuations in wind speed and direction, can have a significant impact on turbine performance and loading. The most common indicator of turbulence is the standard deviation (σ) of the wind speed calculated from 1- or 2-s samples over a 10-min recording interval. Dividing this value by the mean wind speed for the same interval gives the turbulence intensity (TI):

$$\mathrm{TI} = \frac{\sigma}{v} \tag{10.8}$$

where

σ = the standard deviation of wind speed for the recording interval;

v = the mean wind speed for the recording interval.

The mean TI generally decreases with increasing wind speed up to about 7–10 m/s, above which it is relatively constant. TI values above 10 m/s typically range from less than 0.10 in relatively flat terrain with few trees or other obstacles to more than 0.25

in forested, steep terrain. Along with the mean wind speed and air density, the TI at 15 m/s enables a preliminary determination of the suitability of a turbine model for the project site. The final determination is usually made by the manufacturer, who is responsible for the warranty, and may take into account the frequency distribution of turbulence as well as turbulence generated by upstream turbines. (See Chapter 16.)

10.1.5 Wind Power Density

WPD is defined as the flux of kinetic energy in the wind per unit cross-sectional area. Combining the site's wind speed distribution with air density, it provides an indication of the wind energy production potential of the site. It is calculated in the following way:

$$\text{WPD} = \frac{1}{2N} \sum_{i=1}^{N} \rho_i v_i^3 \ (\text{w/m}^2) \tag{10.9}$$

where
$\quad N =$ the number of records in the period;
$\quad \rho_i =$ the air density (kg/m^3);
$\quad v_i =$ the wind speed for record i (m/s).

The air density in this equation must be calculated from other information, as described in the following section.

Note that the cubic equation must be evaluated for each record and then summed, as shown, rather than being applied to the mean wind speed for all records. This is because above-average wind speeds contribute much more to WPD than do below-average speeds, thanks to the cubic exponent. Even then, the WPD estimate is not exact since it ignores variations in speed within each recording interval. The true WPD is generally a few percent greater than that calculated from this formula. This is usually not important for wind resource assessment, since WPD is not used directly in calculating energy production.

10.1.6 Air Density

The air density depends on temperature and pressure (and thus altitude) and can vary by as much as 10–15% seasonally. If the site pressure is measured, the air density can be calculated from the ideal gas law:

$$\rho = \frac{P}{RT} \ (\text{kg/m}^3) \tag{10.10}$$

where
$\quad P =$ the site air pressure (Pa or N/m^2);
$\quad R =$ the specific gas constant for dry air (287.04 J/kg · K);
$\quad T =$ the air temperature in degrees Kelvin ($^\circ$C + 273.15).

If the site pressure is not available (as is usually the case), the air density can be estimated as a function of the site elevation and temperature, as follows:

$$\rho = \left(\frac{P_0}{RT}\right) e^{\left(\frac{-gz}{RT}\right)} (\text{kg/m}^3) \qquad (10.11)$$

where

P_0 = Standard sea-level atmospheric pressure in Pascal (101,325 Pa)
T = Air temperature (K), $T(\text{K}) = T(^\circ\text{C}) + 273.15$
g = the gravitational constant (9.807 m/s^2)
z = the elevation of the temperature sensor above mean sea level (m).

After substituting the numerical values for P_0, R, and g, the equation becomes

$$\rho = \left(\frac{353.05}{T}\right) e^{-0.03417\frac{z}{T}} (\text{kg/m}^3) \qquad (10.12)$$

While this equation is quite accurate (to within 0.2% at most sites), the error increases with increasing elevation because the air pressure does not follow the exponential function exactly.

10.1.7 Speed Frequency Distribution and Weibull Parameters

The speed frequency distribution is a critical piece of information as it is used directly in estimating the power output of a wind turbine. The frequency distribution represents the number of times in the period of record that the observed speed falls within particular ranges, or bins. The speed bins are typically 0.5 or 1 m/s wide and span at least the range of speeds defined for the turbine power curve, that is, from 0 to 25 m/s and above. It is usually presented in reports as a bar chart, or histogram, covering all directions. In addition, the wind speed frequency distribution by direction is stored in a tabular format, which is used as an input to wind plant design software.

The Weibull distribution is a mathematical function that is often used to represent the wind speed frequency distribution at a site. In the Weibull distribution, the probability density (the probability that the speed will fall in a bin of unit width centered on speed v) is given by the equation:

$$p(v) = \frac{k}{A} \left(\frac{v}{A}\right)^{k-1} e^{-\left(\frac{v}{A}\right)^k} \qquad (10.13)$$

There are two parameters in the Weibull function: A, the scale parameter, which is of dimension speed and is related closely to the mean wind speed, and k, the non-dimensional shape parameter, which controls the width of the distribution. Values of k range from 1 to 3.5, the higher values indicating a narrower frequency distribution (i.e., a steadier, less variable wind). A commonly observed k range is 1.6 to 2.4. Within this range, the mean speed is about 0.89 times the scale factor. Figure 10-3 illustrates Weibull probability density curves for several values of k and constant A.

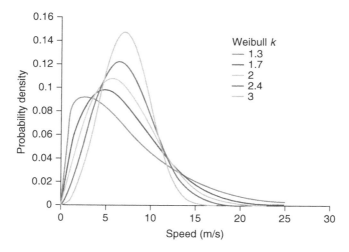

<u>Figure 10-3.</u> Weibull probability density curves for a range of values of k. All curves have the same A: 8.0 m/s. *Source*: AWS Truepower.

It is often handy to refer to the Weibull parameters, particularly k, when characterizing a site's wind resource. It is important to keep in mind, however, that the Weibull curve is, at best, an approximation of the true wind speed frequency distribution. While the real speed distributions at many sites fit a Weibull curve quite well, there are some sites where the fit is poor, as suggested in Figure 10-4. For this reason, *the Weibull curve should never be used in place of the observed speed frequency distribution when estimating energy production*, except in a preliminary way. Many resource analysts choose to ignore it altogether.

10.1.8 Wind Rose

In most projects, it is desirable to space turbines much further apart along the principle wind direction than perpendicular to it to minimize wake interference between the turbines. For this reason, the directional frequency distribution is a key characteristic of the wind resources.

A polar plot displaying the frequency of occurrence by direction is called a *wind rose*. Wind rose plots often display the percentage of time the wind blows in certain speed ranges by dividing each segment of the plot into different color bands. Another type of plot, known as an *energy rose*, displays the percentage of total energy in the wind coming from each direction. Sometimes these plots are combined into one. Wind and energy rose plots are created by sorting the wind data into the desired number of sectors, typically either 12 or 16, and calculating the relevant statistics for each sector.

$$\text{Frequency}(\%) : f_i = 100 \frac{N_i}{N}$$

$$\text{Percentage of total energy} : E_i = 100 \frac{N_i \times \text{WPD}_i}{N \times \text{WPD}}$$

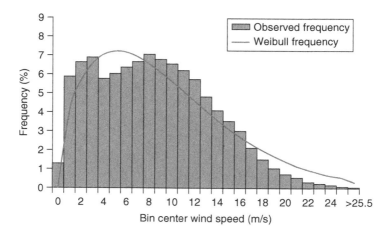

Figure 10-4. An example of an observed speed frequency distribution plotted against a fitted Weibull curve. The observed distribution in this case is bimodal, that is, it has two peaks, and does not follow the Weibull curve very well. At most sites, the Weibull function fits the observed distribution better than this. *Source*: AWS Truepower.

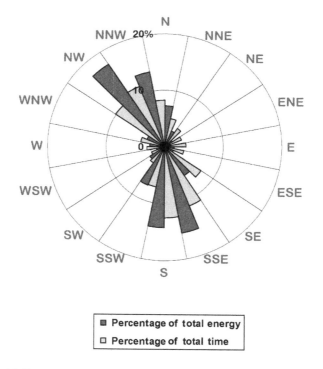

Figure 10-5. Wind and energy rose plot example. *Source:* AWS Truepower.

In these equations, N_i refers to the number of records in direction sector i, N is the total number of records in the data set, WPD_i is the average wind power density for direction sector i, and WPD is the average wind power density for all records. Figure 10-5 contains a typical plot showing both frequency and percent of energy.

10.2   QUESTIONS FOR DISCUSSION

1. Describe some situations in which the annualized mean wind speed could differ significantly from the observed mean speed. Explain why the annualized mean is usually a better representation of the wind resource.

2. Explain the meaning of WPD. Using the Internet, find one or more references that describe the wind resource of a site or region in terms of WPD, and compare the estimates you find with corresponding estimates of mean wind speed. Is there a one-to-one correspondence between WPD and mean speed?

3. From the perspective of a wind project developer, rank the wind resource characteristics described in this chapter from most to least important. Why did you choose this order?

4. Why is it important to maintain a height ratio of 1.5 or greater between two levels for wind shear measurement? Suppose the relative speed error is 1.2% and the two heights are 50 and 40 m. What is the shear contribution to the speed error at 80 m?

5. Name a type of wind speed frequency distribution not represented well by the Weibull function.

6. Suppose the observed wind speed frequency distribution at a site has a k value of 1.3. Would you say the winds at this site are more or less variable than a site with a k value of 2.3? Assume the two sites have the same value of A, 9 m/s. Is a wind speed of 12 m/s more or less likely to occur at the first site compared to the second? Which site is likely to generate more energy?

7. Why do you think it is useful to include more than just the frequency by direction in a wind rose chart? Would you prefer to include mean speed or percent energy by direction? Explain why.

8. The effects of water vapor have been ignored in Equation 10.9. Describe how accounting for water vapor would affect the air density estimate.

9. What impact does air density have on the available kinetic energy in the wind? For the same speed frequency distribution, would a high air density or low air density site produce more energy for the same mean wind speed? Why?

10. Why is knowing the wind rose important for designing the instrumentation on a meteorological mast? Is it also important for designing the turbine layout?

SUGGESTIONS FOR FURTHER READING

Burton T, Sharpe D, Jenkins N, Bossanyi E. Wind energy handbook. West Sussex: John Wiley & Sons; 2001.

Danish Wind Industry Association. Guided Tour on Wind Energy. Available at http://www.heliosat3.de/e-learning/wind-energy/windpowr.pdf.

Manwell JF, McGowan JG, Rogers AL. Wind energy explained: theory, design, and application. 2nd ed. West Sussex: John Wiley & Sons; 2009.

11

ESTIMATING THE RESOURCE AT HUB HEIGHT

Since the height of the top anemometers on many towers is below the hub height of modern, large wind turbines, it is often necessary to extrapolate the wind resource data to the turbine hub height, where the power curve is defined.[1] This is so not only for wind speeds but also for other information such as air density and turbulence intensity (TI). At this stage, the resource analyst begins to depart from what is strictly measured to what must be assumed or modeled. The task requires a careful and often subjective analysis of information about the site, including the local meteorology, topography, and land cover, as well as the data themselves. This chapter discusses methods of carrying out this analysis.

11.1 WIND SPEED

The most widely used method of projecting the wind speed from the height of observation to the turbine hub height is by means of the power law. This was described

---

[1] Although estimating the wind at hub height remains the main goal of resource assessment, attention is increasingly being paid to characterizing the resource from the bottom to the top of the turbine rotor. This is especially important at sites where the shear is highly uncertain or variable. Where such conditions are suspected, extra-tall towers or ground-based remote sensing may be called for. The principles outlined in this chapter nonetheless still apply.

Wind Resource Assessment: A Practical Guide to Developing a Wind Project, First Edition.
Michael Brower et al.
© 2012 John Wiley & Sons, Inc. Published 2012 by John Wiley & Sons, Inc.

briefly in Chapter 10, and the equation is reproduced here:

$$v_2 = v_1 \left(\frac{h_2}{h_1}\right)^a \tag{11.1}$$

As written here, h_1 and h_2 refer to the top anemometer height and hub height, respectively, and the equation has been rearranged so that the known parameters are all on the right. Surprisingly, the power law equation has no basis in meteorological theory, but it has proved highly useful in practice because it fits many wind speed profiles quite well, is simple, and requires just one parameter, the shear exponent, α. An alternative approach more firmly rooted in theory is based on a logarithmic equation. It is discussed later in this chapter.

The key question when applying the power law is what to assume for the shear exponent. It might seem reasonable to use the exponent that was calculated between the first (top) and second heights on the tower, or if the ratio of those two heights is not large enough to obtain an accurate shear value, between the first and third heights. This is in fact a reasonable starting point, but it is not the end of the analysis.

The challenge is to determine whether the shear exponent changes with height above the mast top. In fairly open and flat terrain, it usually does not change very much. But the assumption of constant shear exponent may not hold in other situations, such as where there is dense forest or strong topographic enhancement or blocking of the wind flow, or when the most energetic wind is confined to a relatively shallow layer near the surface, as occurs in thermally driven drainage (katabatic) wind flows. In such situations, adjustments to the mean shear exponent may be warranted.

The following sections discuss several strategies that can be followed to determine what shear adjustment, if any, is warranted.

11.1.1 Direct Measurement

The ideal is to measure the wind speed up to (and even beyond) the turbine hub height. This can be done with either a hub-height mast or a ground-based sodar or lidar system. Strictly speaking, such measurements apply only to the locations at which they were taken. Why not then simply measure the profile to hub height at every mast? Desirable though that would be, there are two reasons why it is not always practical. First, hub-height towers are expensive and difficult to install in some sites. Second, it is often not possible to deploy sodars or lidars right next to an existing mast. This may not matter very much if the site is relatively uniform—open land cover with few trees and little steep terrain—so that the shear is more or less constant across the project area. Then the shear to hub height observed at one location may be applied with some confidence to other locations.

At more complex wind sites, however, the remotely sensed wind profile may not match the profile observed at the site's primary meteorological towers very closely. In that case, one is left to decide how to modify the observed shear at each of the towers to account for the information provided by the hub-height mast or remote sensing systems. One approach is to apply an adjustment to the shear observed at the mast based on the change in the shear observed in the sodar data:

$$\text{Difference method}: \quad \alpha^{(m)}_{h_2 \to h_h} = \alpha^{(m)}_{h_1 \to h_2} + (\alpha^{(s)}_{h_2 \to h_h} - \alpha^{(s)}_{h_1 \to h_2}) \quad (11.2)$$

The superscripts (m) and (s) refer to the shears measured by the mast and sodar (or lidar), respectively. The subscripts $h_1 \to h_2$ and $h_2 \to h_h$ refer to the shears measured between the bottom height (1) and top height (2) of the anemometers on the mast and between the top height and the hub height. Though the approach is reasonable, care must be taken to avoid unrealistic or unreasonable results.

To see how such an adjustment might be applied in practice, consider the following examples.

1. *Simple Case.* Suppose a 60-m tower and a sodar unit are deployed in a relatively open and flat project area. Also suppose that the shear exponent from 40 to 60 m measured at the mast is 0.18 and that the exponent measured by the sodar is 0.20 from 40 to 60 m, decreasing to 0.16 from 60 to 80 m (the presumed hub height). It would be reasonable to assume in this case that the shear at the mast also decreases with height. The difference method would produce a reduction of 0.04 in the exponent at each tower.

2. *Challenging Case.* Suppose a 60-m tower and a sodar unit are deployed in complex, forested terrain. The mast is near the edge of a steep drop-off, whereas the sodar is deployed 100 m back from the edge, in a clearing within deep forest. The shear exponent between 40 and 60 m observed at the mast is 0.25, while the exponent measured by the sodar is 0.40 between 40 and 60 m and 0.20 between 60 and 80 m. Using the difference method, the inferred shear from 60 to 80 m at the mast would be $0.25 + (0.20 - 0.40) = 0.05$. This result seems unrealistically low, however. It is clear that the sodar is not in a good location for interpreting the mast profile. In this case, the reasonable course might be to assume a slight reduction, to perhaps 0.20, in the mast shear to hub height.

As the second example shows, direct measurements of the shear to hub height are not always easy to interpret, nor do they remove all the uncertainty in extrapolating the observed anemometer speeds to hub height. As sodar and lidar technology improves and becomes more widely used, it is likely that the uncertainties in the process will diminish.

When relying on much less than a year of direct measurements to hub height, it is important to consider seasonal variations in the wind shear as well. Where possible, only concurrent data should be compared between towers or between a tower and remote sensing system. For example, the wind shear measured by sodar over a 1-month period should be compared to that measured at a tower over the same period. Because wind shear can vary widely depending on atmospheric conditions, if concurrent data are not available, it may not be possible to use the remotely sensed data at all. In general, the most accurate adjustments requires either a full year, or a statistically representative sample of a full year, of direct measurements from a hub-height tower or remote sensing system.

11.1.2 Displacement Height

One reason the wind shear can vary with height is that the wind flow is displaced above the ground by vegetation, such as a dense, closed-canopy forest. The effective ground level where the wind speed profile reaches zero is then some distance, known as the *displacement height*, above the actual ground level. The displacement height depends on the height and density of the surrounding vegetation and on the distance of the vegetation from the base of the tower. As a rule of thumb, for dense vegetation close to the tower, it is 0.6–0.9 times the vegetation height (1). The shear exponent calculated with respect to the displacement height d is as follows:

$$\alpha_{h_1 \to h_2}^{(d)} = \frac{\log\left(v_2/v_1\right)}{\log[(h_2 - d)/(h_1 - d)]} = \alpha_{h_1 \to h_2}^{(g)} \frac{\log(h_2/h_1)}{\log[(h_2 - d)/(h_1 - d)]} \tag{11.3}$$

The superscripts (d) and (g) denote the shear exponent referenced to the displacement height and ground, respectively. Since the fraction on the right is always less than 1, the shear exponent with respect to the displacement height is always less than the shear exponent with respect to the ground.

Now let us suppose that the exponent relative to the displacement height remains constant with height (this is equivalent to saying that the displacement effect is responsible for all the change in shear with height). Then, the shear with respect to ground between the top anemometer height and the hub height is given by

$$\alpha_{h_2 \to h_h}^{(g)} = \alpha_{h_1 \to h_2}^{(g)} \frac{\log[(h_h - d)/(h_2 - d)]}{\log(h_h/h_2)} \frac{\log(h_2/h_1)}{\log[(h_2 - d)/(h_1 - d)]} \tag{11.4}$$

This equation is valid only for heights greater than the displacement height. The adjusted exponent is always smaller than the observed exponent relative to ground; in other words, the displacement effect causes the shear exponent relative to ground to decrease with height.

As an illustration, suppose the wind shear exponent measured between 40 m (h_1) and 60 m (h_2) is 0.35 and the tower is surrounded by dense, leaf-covered trees that are 15 m tall on average. The analyst estimates a displacement height (d) of roughly two-thirds the tree height, or 10 m. The projected shear exponent from 60 m to the hub height (h_h) of 80 m is

$$0.35 \times \frac{\log(70/50)}{\log(80/60)} \times \frac{\log(60/40)}{\log(50/30)} = 0.325$$

representing a 7% decrease over the observed exponent. Figure 11-1 shows the change in apparent shear relative to the ground for various combinations of observed shear exponent and displacement height for tower heights of 40 and 60 m and a hub height of 80 m.

A directionally weighted displacement height can be calculated if sufficient information is available; site photos can be helpful in this regard. The displacement height

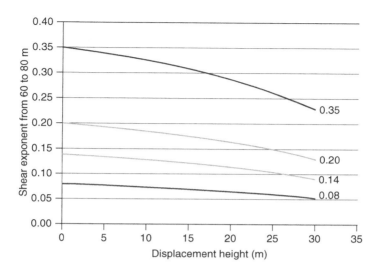

Figure 11-1. Effect of displacement height on the shear exponent between the top height of a mast and the hub height, for different values of observed shear. The observed shear is calculated between 40 and 60 m, and the hub height is assumed to be 80 m for each case. *Source*: AWS Truepower.

should be reduced as the distance between the vegetation and the tower increases. Unfortunately, there are no proven guidelines for estimating this reduction. In the absence of such guidelines, a linear reduction over a distance of 20–50 times the vegetation height is reasonable.

As a practical matter, if the vegetation is no more than a few meters tall, the displacement has little impact and can safely be ignored. It is only in forest with trees more than about 10 m tall that the effect of displacement on the expected shear above the top height of the mast becomes significant.

For towers with at least four measurement heights, it is sometimes possible to dispense with displacement height and fit the observed change in shear directly to the data. A logarithmic function often works well for this purpose. This approach is rarely feasible, however, because few towers have as many as four levels. In addition, if the lowest level is sheltered by trees or obstacles, the fitted curve may not be reliable.

11.1.3 Convergence Height

A characteristic of the atmospheric boundary layer is that the influence of variations in topography and land cover tends to diminish with height. This idea is neatly captured in the concept of a convergence height, which is defined as the height above ground where the wind speed profiles at different points in a project area converge and the wind resource becomes homogeneous.

There is no standard or easily calculated convergence height: it varies greatly depending on the site and atmospheric conditions. In nearly flat, featureless terrain, it may be at virtually ground level, meaning the wind resource varies scarcely at all across the site. In complex terrain and with mixed land cover, the convergence height might be several hundred to thousands of meters above the mean ground elevation.

What is most useful about the convergence height is its corollary: near the ground, wind shear tends to be inversely related to mean speed. This is because the greater the shear is, the more rapidly the speed decreases from the convergence height down to the surface, and therefore, the lower the speed at the height of measurement. (This is implied in Figure 10-1 if one imagines the convergence height is 120 m.)

The presumed inverse relationship between wind shear and speed occurs only if the shear is fairly constant with height, or at least varies with height in a similar way across the project area. This, as we have seen, does not always hold true. Nevertheless, such a relationship is observed surprisingly often and can be used as a tool to inform judgments about the likely change in shear above a particular mast.

For example, suppose a mast exhibits both a high shear and a high mean speed relative to other masts in the project area. It is reasonable to conclude that this shear cannot persist to a great height, as otherwise the speed profile above the mast would diverge from the others. Likewise, a mast with a relatively low speed and low shear can be expected to see an increase in shear with height. A simple scatter plot of shear exponent versus mean speed can be used to identify towers that deserve closer examination (note that the speeds and shears should be for the same heights and time periods). Any outliers should first be examined for poor data quality, incorrect instrument heights, or other problems that could account for the apparent discrepancy. If no such errors are found, the shear exponents for the outlying towers can be adjusted to avoid unrealistic divergence in the speed profiles. The approach is illustrated in Figure 11-2, which is taken from a wind project site in the United States. In this case, the shear at the outlying point circled in red was adjusted downward.

11.1.4 Logarithmic Method

The most commonly used logarithmic expression for wind shear is the following equation:

$$v_2 = v_1 \frac{\log\left(h_2/z_0\right)}{\log\left(h_1/z_0\right)} \tag{11.5}$$

Here, z_0 is the surface roughness length, a parameter linked to the height and density of vegetation and other rough elements surrounding the tower. Strictly speaking, this equation is applicable only when the boundary layer is neutrally buoyant.[2] At times

---

[2]Buoyancy is defined in terms of the adiabatic rate of temperature change with height. This is the rate of change in temperature of a parcel of air that is displaced upward or downward, only due to its change in pressure, with no exchange of heat with the surrounding air. If the actual temperature lapse rate exceeds this critical rate then a parcel of air displaced upward will find itself cooler and heavier than the surrounding air and so will tend to sink back down: it is negatively buoyant or thermally stable. Conversely, if the lapse rate is lower than the critical rate, the atmosphere is said to be thermally unstable and convective mixing results.

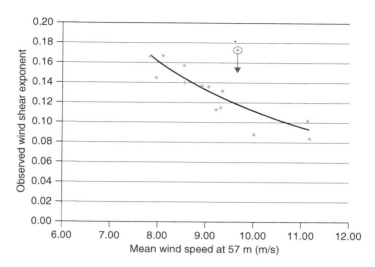

<u>Figure 11-2.</u> Scatter plot of wind shear versus mean wind speed at 57 m, showing an outlying mast where the shear is higher than expected for the speed. In this case, a reduction in the shear could reasonably be applied. *Source:* AWS Truepower.

of thermal stability (negative buoyancy), a more complicated form with an additional parameter, the stability length, should be used. The effect of this additional parameter is to increase the shear. Unfortunately, the stability length must be estimated from temperature data at multiple heights, which is rarely available.

Out of convenience, therefore, most analysts rely only on the neutral form of the equation. This is important because if the roughness is determined from the vegetation (or other land cover) characteristics alone, there is a tendency to substantially underestimate the wind shear, especially in temperate climates like that of North America. Thus, z_0 must usually be treated as an empirical parameter that is fit to the data, much like the shear exponent. Table 11-1 indicates values of z_0 corresponding to values of α for shears calculated between 40 and 60 m height.

An important question is how the two methods compare when projecting speeds to hub height. Maintaining a constant value of z_0 is equivalent to reducing the shear exponent with increasing height. In going from 60 m to 80 m, α is reduced by about 5–11% in the shear range 0.14–0.35. The impact on the hub-height mean speed ranges from about $-0.2\%(\alpha = 0.14)$ to $-1.1\%(\alpha = 0.35)$.

Thus, the logarithmic approach is the more conservative. However, while there are certainly many sites where the shear exponent decreases with height, there are many others where it holds steady or increases. For example, in the Great Plains of the United States, the well-known nocturnal jet phenomenon (caused by a decoupling of the lower atmosphere from surface roughness under stable nighttime conditions) often produces an increasing shear exponent with height. The limited research that has been published comparing the power law and logarithmic methods does not establish clearly which is the more accurate under most circumstances.

Table 11-1. Equivalence of z_0 and α
for neutrally buoyant conditions and
heights of 40 and 60 m

α	z_0
0.08	0.0002
0.14	0.039
0.2	0.33
0.35	2.8

Overall, the differences between the power law and logarithmic approaches are small, and which one is used is largely a matter of the resource analyst's preference. For the remainder of this chapter, only the power law method is discussed.

11.2 TIME-VARYING SPEEDS AND SPEED DISTRIBUTIONS

While the time-averaged shear exponent is a convenient parameter for characterizing the wind resource at a site, it is not appropriate for scaling a time series of wind speed data or a speed frequency distribution. The reason is that it overlooks the wide variation in shear, especially its dependence on direction, time of day, and time of year. The plot in Figure 11-3 illustrates a common pattern of variation of shear with time of day, showing a minimum during daytime hours when the atmospheric boundary layer is well mixed and a maximum at night under thermally stable conditions. Relying on an average shear can introduce errors in the speed distribution at hub height, with an impact on energy production estimates.

One approach to this problem is to calculate a shear exponent for each time interval (e.g., 10 min) and to use that exponent to extrapolate the top anemometer speed to hub height. A potential drawback of this "instantaneous shear" method is that extreme (albeit valid) shear values can occasionally occur in the record, which may not persist to hub height. These extreme values can produce unrealistically high or low hub-height speeds, but they are generally rare. A more common problem is that shear values are available only for those records for which both the upper and lower sensors have valid data. The method should not be applied to substituted data. Whether this is a serious problem depends on the amount of data substitution that has occurred.

As an alternative, many analysts choose to bin the speeds by direction, time of day, or time of year, or a combination of these, and to calculate the mean speed and average shear for each bin. The 10-min wind speeds can then be extrapolated to hub height using the shear for the appropriate bin for each record. This resolves the problem of extrapolating substituted data, while preserving at least part of the full variation of shear. Care must be taken to handle bins with few or no points in an appropriate way, such as averaging speeds and shears from adjacent bins. When using

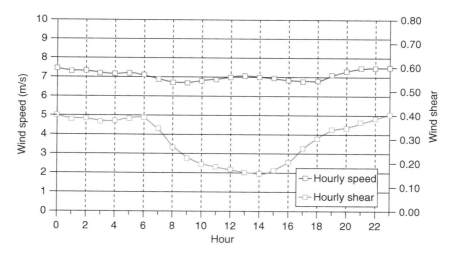

<u>Figure 11-3.</u> A typical pattern of diurnal variation of mean wind speed and shear. In this example, the speed is observed at 60 m, and the shear is calculated between 40 m and 60 m. *Source:* AWS Truepower.

this method, it is almost always important to bin by time of day to capture the effect of diurnal variations of atmospheric stability.

Where the shear is believed to change with height, using instantaneous or binned shear exponents can introduce an additional challenge since each shear value must in principle be adjusted separately, and there is rarely enough information to do this with confidence. An alternative is to rely on the time-averaged shear, adjusted if necessary for displacement and other effects, to establish the mean wind speed at hub height. Any difference between this mean and that derived from the extrapolated time series or speed distribution can be resolved by rescaling the extrapolated data so that the means match. The rescaling is done by multiplying each speed value by the ratio of the expected hub-height mean speed and the mean of the extrapolated data. This method, while not ideal, represents a commonsense compromise between the simplicity and transparency of using a time-averaged shear and the greater accuracy of speed distributions derived from binned or time-varying shears.

11.3 OTHER PARAMETERS

Three other wind resource characteristics, the wind direction, air density, and TI, must also be projected to hub height for estimating turbine and plant power production.

11.3.1 Wind Direction

It is generally assumed that the wind direction is constant with height above the top anemometer. This is not strictly true even in principle, as the interaction of the earth's rotation with frictional and pressure forces tends to produce a rotation of the wind vector with increasing height above ground. This effect, however, is quite small, so a constant direction is usually a safe assumption. Therefore, the directions recorded at the top anemometer vane (with substitutions as needed) are generally projected to hub height with no alteration.

There are sites where, because of the influence of terrain or strong temperature gradients, substantial directional changes with height (veer) are frequently observed. Such shifts can reduce turbine power production since the wind vector may not be perpendicular to the rotor throughout the rotor plane. Remote sensing using sodar or lidar can detect these situations.

11.3.2   Air Density

Two factors affect how the air density varies with height: the pressure (or elevation) and air temperature. The air temperature is usually extrapolated from the thermometer height to the hub height using the temperature lapse rate of the standard atmosphere of 6.5°C (K) per 1000 m. For a thermometer height of 3 m and a hub height of 80 m, this represents a drop in temperature of about 0.5°C. Applying the change in temperature and height to the density equation (Eq. 10.12), the effect is to decrease the air density by 0.6–0.8% (independent of elevation).

11.3.3   Turbulence Intensity

The TI at hub height is normally estimated by assuming that the 10-min speeds increase with height above the top anemometers, while their standard deviations remain constant. Thus, the TI for each record decreases with height, and the speed-averaged TI follows suit. The mean TI for the 15 m/s bin—a standard reference parameter for determining if a particular turbine model is suitable for a site—may change slightly, if at all.

11.4   QUESTIONS FOR DISCUSSION

1. What are the advantages and disadvantages of using hub-height monitoring masts or remote sensing systems to measure the hub-height wind resource? How might the answer differ in simple and complex terrain?

2. Define what the convergence height is and how its presence can be applied for the shear analysis of a site. When does the convergence height corollary not hold true?

3. Name three methods for scaling a time series of wind speed data or a wind speed frequency distribution to hub height. Which method do you think is best?

4. Which two factors affect how air density varies with height?

5. A meteorological mast measures average wind speeds of 7.1 m/s at a height of 58.5 m and 6.2 m/s at 32.2 m.

- Using Equation 10.6, calculate the shear exponent.
- Assume this exponent remains constant with height. From the power law, extrapolate the mean speed from 58.5 to 80 m.
- Using Equation 11.5, in a spreadsheet or with a calculator, find the value of z_0 that gives the correct speed at 58.5 m. For this purpose set $h_1 = 32.2$, $h_2 = 58.5$, and $v_1 = 6.2$. (This will take some trial and error with different values of z_0. You can use the Microsoft Excel Solver function if you wish.) Looking at Table 11-1, does the value you derive make sense?
- Using the same value of z_0, project the mean speed at 58.5 m to 80 m. For this purpose, set $h_1 = 58.5, h_2 = 80.0$, and $v_1 = 7.1$.
- Compare the results from the second and fourth points of this problem. How much do the extrapolated speeds differ in percentage terms? Relative to the logarithmic approach, does the power law produce a higher or lower shear when extrapolating above the top measurement height?

6. A 60-m meteorological mast has wind speed monitoring heights of 30, 40, and 60 m. The average wind speeds at each level are 5.3, 5.7, and 6.4 m/s, respectively.
   - Using the power law, calculate the observed shear exponent between the following pairs of levels: 30 and 40 m, 40 and 60 m, and 30 and 60 m.
   - Assuming the speeds and heights are all known to the same degree of certainty, which of these shear values would you say is the most precise?
   - Assume the heights are perfectly known and the speed ratio between any two heights has an uncertainty of 1.5% (i.e., in Eq. 10.7, $\varepsilon = 0.015$). What is the uncertainty of each shear exponent you calculated? Do you think the variation in shear between heights is statistically significant? Which shear exponent do you think should be used to extrapolate the 60-m wind speeds to 80 m, and why?

7. A 60-m meteorological mast is deployed in forested terrain with an approximate 7-m displacement height in all directions. The shear exponent observed at the mast is 0.32 between 40 and 60 m. Assume the shear exponent with respect to the displacement height remains constant with height. What would be the observed shear with respect to ground between the top anemometer height and an 80-m hub height?

8. A meteorological mast has a thermometer installed at a height of 5 m. By how many degree Celsius would you expect the average temperature to decrease between 5 and 80 m?

9. A 50-m meteorological mast observes a 10-min average wind speed of 6.8 m/s and a corresponding standard deviation of 1.1 m/s. What is TI? If the concurrent wind speed at a hub height of 80 m is projected to be 7.5 m/s, what would be a good assumption for the hub-height TI?

REFERENCE

1. Garratt JR. The atmospheric boundary layer. Cambridge, UK: Cambridge University Press; 1992. p. 290.

SUGGESTIONS FOR FURTHER READING

Feuquay LF. Validation study of the use of wind shear exponents in extrapolating wind speeds for wind resource estimations. In: 13th Symposium on Meteorological Observations and Instrumentation; USA: 2005.

Garratt JR. The atmospheric boundary layer. UK: Cambridge University Press; 1992.

Lubitz WD. Accuracy of vertically extrapolating meteorological tower wind speed measurements. Conference Proceedings: CanWEA; Winnipeg, Manitoba, Canada; 2006. Available at http://www.soe.uoguelph.ca/webfiles/wlubitz/Lubitz_CanWEA_2006.pdf. (Accessed 2012).

Ray ML, Rogers AL, McGowan JG. Analysis of wind shear models and trends in different terrains. In: Conference Proceedings: AWEA Windpower; Pittsburgh, Pennsylvania, USA; 2006. Available at http://ceere.org/rerl/publications/published/2006/AWEA%202006%20 Wind%20Shear.pdf. (Accessed 2012).

THE CLIMATE ADJUSTMENT PROCESS

The last major step in characterizing the wind resource at a wind monitoring station, before extrapolating it to the wind turbine locations, is to adjust the observed wind climate to the historical norm. Average wind speeds can vary substantially from the norm even over periods of a year or longer. Typically, the uncertainty in the long-term mean wind speed based on a year of measurement alone is about 3–6%,[1] corresponding to perhaps 5–10% in the mean wind plant production—a significant factor when assessing the risk of financing a wind project. Reducing this uncertainty is the primary goal of the climate adjustment process.

The leading method for performing climate adjustments is popularly called *MCP*, which stands for measure, correlate, predict. The wind resource is measured at a site (sometimes called the *target*) over a period ranging from several months to several years. The observed winds at the target site are correlated with those recorded at a long-term reference, such as an airport weather station, and a relationship between

---

[1] On the basis of an analysis by AWS Truepower of data from first-order meteorological stations in North America. At some locations, the interannual variability may be larger or smaller than this range.

Wind Resource Assessment: A Practical Guide to Developing a Wind Project, First Edition.
Michael Brower et al.
© 2012 John Wiley & Sons, Inc. Published 2012 by John Wiley & Sons, Inc.

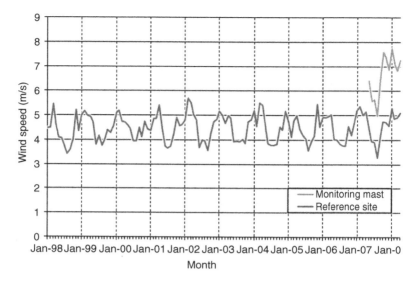

<u>Figure 12-1.</u> Observed wind speeds from a monitoring mast and reference station. *Source: AWS Truepower.*

them is established. Then, the much longer historical record from the reference is applied to this relationship to predict the long-term mean wind resource at the target site. An example of the relationship between the periods of record for a reference site and a monitoring mast is provided in Figure 12-1.

That is how the process is supposed to unfold, and often it does so successfully. However, complications sometimes arise, and the accuracy of the resulting long-term wind resource estimates depends on how they are handled.

This chapter addresses (i) the assumptions underlying MCP; (ii) the requirements that must be met for successful MCP; (iii) the data sources most widely used for MCP, including their advantages and pitfalls; and (iv) various methods of relating winds observed at the target and reference and of predicting the long-term wind resource.

12.1   IS THE WIND CLIMATE STABLE?

The key assumption underlying all MCP methods is that the wind resource in the future will be similar to what it has been in the past; in other words, the wind climate is stable. In this age of climate change, it is reasonable to ask whether this assumption holds true and what its implications might be for the accuracy of energy production forecasts. Even in the absence of climate change linked to greenhouse gas emissions, the possibility of other sources of change in a site's wind climate, including cyclical weather patterns such as the El Niño Southern Oscillation (ENSO) and local factors such as urbanization and changing vegetation cover, must be considered.

12.1.1 Historical Evidence

Historical evidence concerning the long-term stability of the wind climate is mixed. A key problem is that there are few wind monitoring stations where the wind has been measured continuously at the same height and at the same location, with consistent measurement protocols, and using the same instrument or instrument type, for more than 10–15 years. The dearth of truly homogeneous long-term data sets, coupled with the effects of normal short-term climatic fluctuations, makes it difficult to reliably detect trends caused by long-term climate change.

Nonetheless, various researchers have attempted to detect trends in the wind over time. In one of the most comprehensive studies, a group of European researchers reviewed observations from hundreds of surface weather stations in the northern hemisphere and from two types of modeled global weather information known as *reanalysis data* (1). Almost universally, the surface weather stations show a significant downward speed trend amounting to 5–15% over 30 years. The two reanalysis data sets, however, show much smaller trends that vary from positive to negative depending on the region.

Which picture is correct? The authors ascribe the trends in surface observations in part to changes in surface roughness caused by tree growth. While this is plausible, it is also possible that many of the measurements have been affected by uneven maintenance practices and changes in instrumentation, height, and location, as well as the effects of encroaching urbanization, not all of which could be removed by the QC procedures implemented by the researchers. Most such issues have no bearing on wind energy resources. On the other hand, the global atmospheric models that generate reanalysis data are driven mainly by observations from weather balloons, satellites, aircraft, and other platforms. Although they have their own issues (as discussed below), the reanalysis data sets are at least partly insulated from the inhomogeneities afflicting surface wind observations and thus may provide a truer picture of long-term trends.

Looking more closely at specific countries or regions can be instructive. One study, for example, identified a significant decline in mean wind speeds recorded by surface weather stations in the United States since 1973 (2). As the authors state, it seems likely that the advent of the Automated Surface Observing System (ASOS) in the mid-1990s, as well as other changes such as urbanization and tree growth, is responsible for much of this decrease. They note, "We did not attempt to correct for these inhomogeneities, but their presence strongly argues for use of the other data sets..." By and large, the other data sets they mention, mainly reanalysis data, do not support the observed decline in surface wind speeds nor do long-term observations for the same region from weather balloons (3).

On balance, there is little reliable evidence to date that wind resources have either increased or decreased significantly in the past several decades in most areas of the world as a result of climate change. Considering the uncertainties in the data, any changes that have occurred are probably below the level of confident detection.

12.1.2   Prospects for Change in the Future

What about the future? Although the results of research are far from definitive, overall, they point to a probable decrease in wind resources in the middle latitudes of the northern hemisphere over the next 50–100 years. This is consistent with the fact that certain large-scale temperature gradients that drive atmospheric circulation patterns are likely to diminish in a warming world. In particular, the polar regions are likely to warm more than the equatorial regions, thus probably diminishing the strength of the circumpolar jet streams and shifting them poleward. Some regions may experience an increase in wind resources, however.

Following are some examples of research published so far.

- A study drawing on the results of two general circulation models (GCMs), which are global weather models used by climate researchers to predict climate change, under a single scenario of future greenhouse gas levels, suggested that mean annual wind speeds over the lower 48 states of the United States could decrease from 1.0% to 3.2% by 2050 and from 1.4% to 4.5% by 2100, compared to a 1948–1975 baseline. The two models disagree strongly over the magnitude of the reduction, indicating substantial uncertainty in the conclusion (4).

- Researchers applied a statistical "downscaling" method to four GCMs under two greenhouse gas scenarios and projected the effects of climate change at five weather stations in the Northwestern United States. They found that mean wind speeds at these stations could decrease by amounts ranging up to 10%, depending on the station and the time of year, with the greatest reductions occurring in summer at most sites. The results were fairly consistent across models and scenarios (5).

- A high resolution numerical weather prediction (NWP) model was used to downscale a single GCM and greenhouse gas scenario over southern California. This study found a pattern of both moderate increases and decreases in mean wind speed in 2041–2060 compared to 1980–1999. Unlike other studies, this one looked specifically at an area where wind projects are operating: the Tehachapi Pass. A 2–4% decrease in the mean annual wind speed was predicted where the wind projects are concentrated. Most of this decrease occurred during fall to winter; relatively little change was forecast for the main power-producing months of April to August (6).

- A study drawing on the results of eight climate models over China concluded that increasing temperatures would cause decreasing wind power density. As cited in the study, two of the climate models predict a reduction of approximately 14% in wind power density (representing roughly a 5% reduction in mean speed) over the coming century (7).

Given these findings, it seems likely that any changes in the wind climate over the time horizon of wind project investments, up to 25 years, will be modest. Even a 5% decrease in the mean annual wind speed over 50 years, if it occurred in a linear fashion, would result in only a 0.5% decrease in the average wind speed over the first 10 years of a wind project's life. More likely, as some studies have indicated, the pace of change will initially be smaller. For this reason, and for the time being, the "climate change risk" for wind project investments appears to be small compared to other sources of uncertainty in wind resource assessment.

12.1.3 Other Factors That May Affect the Local Wind Climate

Other factors besides climate change linked to greenhouse gases could alter the future wind climate at a project site. Among these are cyclical weather patterns, including the famous ENSO as well as less-well-known phenomena such as the North Atlantic Oscillation (NAO) and the Pacific Decadal Oscillation (PDO). ENSO events, which are especially influential, last from 6 months to 2 years and occur in cycles of roughly every 4–6 years. Although ENSO's root causes are in some dispute, the phenomenon is closely tied to wind-driven variations, or anomalies, in surface water temperatures in the eastern Pacific Ocean. These temperature anomalies have a substantial impact on weather patterns around the Pacific Rim and throughout the world.

To minimize the effects of such cyclical patterns on the outcome of MCP, the reference station record should span at least two or three oscillation periods. For ENSO, that means around 10–15 years, a feasible time horizon for most MCP studies in countries with reliable reference data, but a challenging one in other countries. Because of the problems already noted with older reference data sets, it is far more difficult to correct for the PDO, with a period of 20–30 years, and other very long-period phenomena. Considering that the period of these oscillations is comparable to or exceeds the time horizon of wind plant investments, and that their behavior cannot be reliably forecast, it is questionable whether such corrections should even be attempted.

Changes in land cover around a site, especially tree growth or clearing, can also alter the wind climate. Considering only the displacement effect of a forest very near the turbines, a typical tree growth rate of 0.2 m per year could reduce the mean wind resource at 80 m height by perhaps 0.5% by the end of the first 10 years of a project's life. This impact is comparable in magnitude to possible trends from global climate change, but unlike climate change, it should be easily forecast based on assessments of the forest condition. Conversely, clearing of 10 m of trees around the turbines could *increase* the available wind resource by up to about 2%, depending on the size of the clearing. Both types of impact could be larger if the changes occurred over a much wider area, up to several kilometers, surrounding the project.

12.2 REQUIREMENTS FOR ACCURATE MCP

Assuming the wind climate is stable, three key requirements must be met for MCP to produce a reliable result:

- *The site and reference station must be in substantially the same wind climate.* This means that variations in wind speed at each location should be well correlated in time. The correlation can be assessed qualitatively by plotting a time series of observed wind speeds for both the target and reference stations. A quantitative measure such as the Pearson correlation coefficient (r) can also be used. The square of the correlation coefficient, r^2, can be thought of as the fraction of the variation in the values of one variable that can be explained by a linear equation with another variable.
- *The target and reference stations must have homogeneous wind speed records.* A wind speed record is said to be homogeneous if the measurements have been taken continuously at the same location and height, with the same or equivalent instrumentation. In the case of the reference station, its record should be substantially longer than, and overlap with, that of the target site.
- *The concurrent target–reference period should capture seasonal variations in the relationship.* In practice, this means at least nine continuous months, and preferably a year or more.

It will become clear in the following sections that the first two requirements, in particular, are not always easy to satisfy.

12.2.1 Correlation

When a wind project site is in flat, open terrain, it is often easy to find a weather station in the vicinity that experiences much the same wind climate. In regions of greater complexity, however, it is not uncommon for the available reference stations to be in quite different wind climates. For example, the project site may be on an exposed ridge or mountain top, while the nearest reference stations are all in sheltered valleys; or the project site may be near a coastline, while the available reference stations are well inland. The result can be relatively poor correlations between the target site and reference station (Fig. 12-2).

The weaker the correlation with the reference station, the larger the uncertainty in the adjusted long-term wind resource at the target site. Assuming normally distributed annual wind speed fluctuations and a homogeneous reference station data record, the following simple equation approximates the overall uncertainty in the long-term mean wind speed as a function of the correlation coefficient, r^2:

$$\sigma = \sqrt{\frac{r^2}{N_R}\sigma_R^2 + \frac{1 - r^2}{N_T}\sigma_T^2} \tag{12.1}$$

Here, σ_R and σ_T are the standard deviations of the annual mean wind speeds of the reference and target sites, respectively, as a percentage of the mean. Often, the same standard deviation is assumed for both sites. A typical range based on North American data is 3–6%, though it may be larger or smaller in other regions. N_R is the number of years of reference data, and N_T is the number of years of concurrent reference and target data. Because of seasonal effects, this equation should not be used if $N_T < 1$.

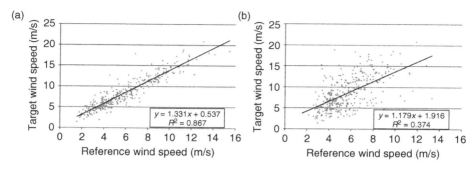

Figure 12-2. Typical scatter plots of target and reference wind speeds. (a) A relatively high correlation, indicating that the two sites experience very similar wind climates. (b) A relatively poor correlation. *Source:* AWS Truepower.

Figure 12-3 plots this equation as a function of r^2 for a range of values of σ. One year of concurrent reference–target data is assumed. Consider the middle curve. When there is no correlation ($r^2 = 0$), the error margin simply equals the annual variability, in this case 4%. For midrange values of r^2, the uncertainty is reduced by one-fourth, to about 3%. If the correlation is very high, the uncertainty is reduced by nearly 70%, to 1.3%. As Figure 12-3 suggests, there is usually no point in employing a reference station with less than a 50% r^2 value; many resource analysts do not consider stations with values of r^2 below 60–70%.

An important question is what averaging interval should be applied to the wind speeds when using the MCP process. The optimal averaging interval for MCP is

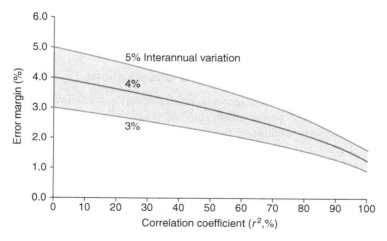

Figure 12-3. The approximate uncertainty margin in the estimated long-term mean wind speed at a site, assuming 1 year of on-site data and 10 years of reference data, as a function of the r^2 coefficient between them and of the interannual wind speed variation σ. *Source:* AWS Truepower.

related to the timescale at which wind fluctuations may be experienced *simultaneously* by the reference and target sites. If the interval is too short, then a large proportion of the speed fluctuations may contain no useful information about the relationship between the two stations; they are just noise. If the interval is too long, then important information about the relationship may be lost.

The optimal time interval is, in turn, related to the size of typical weather disturbances and their rates of motion. As a rule of thumb, the duration of a wind "event," whether it is a gust occurring in a matter of seconds or a sustained period of high winds lasting several days, approximately equals the size of the associated weather disturbance (which may range from a small turbulent eddy to a large storm system or front) divided by its speed relative to the observer. A wind fluctuation cannot occur simultaneously at two points unless both are within the realm of influence of the same disturbance. Thus, the shortest timescale over which correlated fluctuations can occur, Δt (s), is, very approximately, the distance between the target and reference stations, D (m), divided by the typical or average background wind speed, v (m/s):

$$\Delta t \approx \frac{D}{v} \tag{12.2}$$

Suppose the typical mean wind speed is 7 m/s. Then the shortest reasonable timescale to correlate stations that are, say, 100 km apart is about 14,285 s, or 4 h.

As a general guideline, when the reference is an ordinary surface weather station located some distance away from the target tower, daily averaging serves well. This has the advantage that it is simple to apply, and it reduces the influence of differences in diurnal wind speed patterns related to tower height and station location (which can also introduce noise in the correlation). The only time a shorter interval such as 1 h or 10 min might reasonably be used is when the reference station is within a few kilometers of the target site. This occurs most often when secondary masts are correlated with a primary mast within the project area.

12.2.2 Homogeneous Wind Speed Observations

The requirement for a long, homogeneous reference data record can also be difficult to meet. One problem is that measurement standards change from time to time as national weather agencies seek to improve their measurement technology and data products. Unfortunately, this runs counter to the wind industry's interest in having consistent, long-term wind data sets. In the United States, for example, almost all leading weather stations were converted to the ASOS standard in the middle to late 1990s and early 2000s. In the process, tower heights changed from (typically) 6.1 to 10 m, many towers were moved, and the previous manual recording technology was replaced by automated digital equipment. The result was a substantial discontinuity in the recorded wind speeds, which rendered data collected before ASOS effectively useless for MCP.

More recently, cup anemometers, which have been the standard for many decades at US weather stations, were replaced in the United States by ultrasonic "ice-free wind", or IFW, anemometers, resulting in another disruption in the continuity of wind speeds.

Fortunately, the impact of this change was not as severe as that which occurred with the deployment of ASOS. Since there has been no change to the monitoring heights, locations, or data collection practices, adjustment factors can be applied to compensate for the effects of the anemometer change at most ASOS stations. However, there is additional uncertainty associated with this adjustment, which must be considered before relying on the pre-IFW data from these stations.

Similar changes in measurement equipment, tower height, and location have occurred in most countries. Unfortunately, these changes are often not well documented, which leaves the analyst to investigate the history of each reference station— a potentially time-consuming task.

Another challenge is changing site conditions around reference stations, which can create trends in observed wind speeds that have nothing to do with the general wind climate. Figure 12-4 shows an extreme case: a downward trend in observed speeds starting around the middle of the twentieth century at the Blue Hill Observatory in Massachusetts, USA. This trend, which does not appear in other weather records in the region, is at least partially attributable to the regrowth of previously cleared forest around Great Blue Hill, as suggested by the photographs.

In the absence of significant trends or discontinuities, the uncertainty in the long-term mean wind speed derived through MCP should decrease as the length of the reference station's record increases. This is evident in Equation 12.1: the longer the reference data period, N_R, the better. In most real-world situations, however, the benefit of going beyond about 10–15 years of reference data is limited. Figure 12-5 shows the uncertainty for a range of values of r^2 (from 0.45 to 0.95) and N_R (from 1 to 30 years) based on the same equation and assuming $N_T = 1$ and $\sigma_R = \sigma_T = 4\%$. The two dashed curves mark the points where 80% (left-hand curve) and 90% (right-hand curve) of the maximum possible reduction in uncertainty is achieved. For all reasonable values of r^2, 80% of the maximum benefit is reached with less than 10 years of reference data. For $r^2 \leq 0.85$, 90% of the benefit is reached with less than 17 years of data.

The presence of trends or discontinuities in the reference data, whether artifacts of changing site conditions or measurement techniques or real manifestations of climate change, can have a pernicious effect on the accuracy of MCP. Suppose there is a linear trend in the reference wind speed. If the trend does not reflect a real change in the wind climate (perhaps trees are growing around the station or the anemometer has been slowing down because of wear in the bearings), then the adjusted long-term mean wind speed will tend to be biased by an amount ε that depends on the slope of the trend line and the length of the reference data record:

$$\varepsilon \approx -\frac{N}{2}s \tag{12.3}$$

where s is the trend slope in percent per year and N is the number of years in the reference data record. (In this equation, 1 year of overlapping on-site and reference data and perfect correlation between them are assumed.) Thus, where false trends are

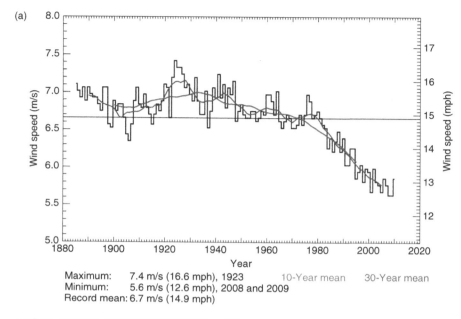

Figure 12-4, (a) Annual average wind speeds and their 10-year and 30-year rolling averages recorded at the Blue Hill Observatory outside Boston from 1885–2010. (b) Photographs of the Blue Hill Observatory taken in 1886 (left) and present day (right). *Source:* Blue Hill Observatory (www.bluehill.org).

present, the magnitude of the potential bias *increases* with the length of the reference period. The problem is compounded if the trend is caused by a real and persistent change in climate; in other words, if the assumption of climate stability is violated. Then, the bias resulting from the ordinary MCP process may be even larger if the trend continues in the future.

Thus, MCP, while still a mainstay of wind resource assessment, cannot remove all uncertainty in the long-term wind climate and must be applied with considerable caution. The following practical guidelines are offered.

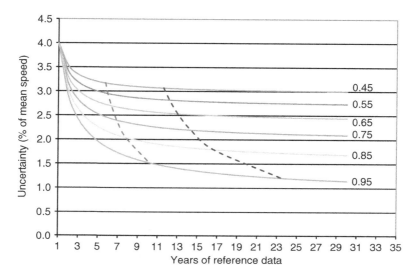

Figure 12-5. Plot of the statistical uncertainty in the long-term mean annual wind speed at the target site as a function of number of years of reference data and for different values of r^2. The two dashed curves show the number of years required to achieve 80% (left dashed curve) and 90% (right dashed curve) of the maximum possible reduction in uncertainty compared to no MCP. The curves are derived from Equation 12.1, assuming 4% interannual wind speed variation, 1 year of overlapping reference and on-site data, and no significant trends or discontinuities in the reference data set. *Source:* AWS Truepower.

- The net should be cast widely for potential reference stations and data sources. The more data sets are available for analysis, the easier it is to detect inhomogeneities.
- The data recovery at the reference stations should be high and consistent over time. Long gaps or significant changes in the data recovery render the data homogeneity suspect.
- The available documentation for each station should be examined carefully to determine whether its instruments, tower height, location, or measurement protocols have changed. The reference period should be the most recent period for which conditions at the station have remained substantially the same.
- The reference data for each station should be assessed visually and statistical tests applied where appropriate to detect trends or inhomogeneities larger than can easily be explained by normal fluctuations.
- The resource analyst should be wary of reference data records extending back more than 15 years, even if there has been no documented change in the station equipment or protocols. There is little confidence that measurements (or modeled data sets such as reanalysis data) as old as this are consistent with more recent data, and there is a risk of introducing significant errors in the MCP process.

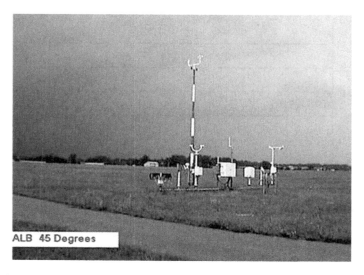

Figure 12-6. The ASOS station at Albany Airport, Albany, New York, USA. *Source:* National Oceanic and Atmospheric Administration (NOAA).

- The analyst should be wary of MCP adjustments, especially upward adjustments, of more than 4% based on a year or more of on-site data, as they may reflect problems in the reference data. Observed trends should be confirmed with at least two or three independent sources of reference data.

12.3 SOURCES OF REFERENCE DATA

Most sources of reference data fall into one of four general categories: tall towers instrumented for wind resource assessment, surface weather stations, rawinsonde stations, and modeled data sets. This section discusses some of the pros and cons of each source of reference data and provides guidance on how the data should be used.

12.3.1 Tall Towers Instrumented for Wind Resource Assessment

It is unusual, but not impossible, to obtain data from tall towers with a sufficiently long record to be useful as reference data sets for MCP. So long as the data prove to be well correlated with the target site and homogeneous through time, they can be an excellent reference. The analyst should be aware of gaps in the data record, as well as possible changes in the anemometers and their mountings and heights. The possibility that wind turbines may have been installed near the tower (a distinct risk given that many such towers are in areas of good wind resource) should also be verified. If the nearest upwind turbine is closer than about 20 rotor diameters to the mast, then the data may not be usable. It is also wise to mistrust published data summaries and preprocessed data files. Whenever possible, the raw data from the towers should be obtained so the analyst can carry out his or her own QC.

12.3.2 Surface Weather Stations

The mainstay of MCP remains surface weather stations. In the United States, ASOS stations are generally preferred for MCP because they are well documented, and, with the exception of the replacement of cup anemometers by ultrasonic IFW anemometers, their instrumentation, maintenance, and data-recording protocols have remained consistent since the ASOS installation date. There are approximately 900 ASOS stations in the United States. Although their geographic distribution is uneven, they provide reasonably good coverage in most parts of the country. Similar networks exist in most other countries, although data quality and availability vary.

Data are reported hourly at most stations. Each speed value is not a true average for the hour but generally represents a 2-min average recorded some minutes before the top of the hour. Thus, even if a station is very near a wind project site, a good correlation of hourly wind speeds should not be expected, and daily means should be employed instead.

Although many surface stations provide a consistent picture of wind speed trends in their respective regions, numerous issues can create inhomogeneities. Three common problems are encroaching urbanization, land clearing (e.g., deforestation), and tree growth. Surface stations located in built-up areas are especially suspect. The analyst should also be alert for stations that may be poorly or inconsistently maintained. These are often apparent because of widely varying rates of data recovery or frequent or long gaps in the data record. Regardless of the situation, the homogeneity of each station's data record must be evaluated case by case, usually by comparing trends from different stations in the same region. Plotting the ratio of monthly or annual mean wind speeds for different pairs of reference stations can be a useful tool for spotting suspicious trends. Where two apparently reliable stations exhibit different trends, one upward and the other downward, and no other stations are available for comparison, it is usually the safer course to reject the downward-trending station because most problems with surface stations are likely to cause a decrease in speed.

12.3.3 Rawinsonde Stations

Data from instrumented weather balloons (Fig. 12-7), known as *rawinsondes*, can sometimes be useful for MCP. One advantage of rawinsonde observations is that they are generally taken well above the land surface (at both fixed and variable heights defined by atmospheric pressure) and so are largely insulated from changes in land cover. Another advantage is that they can be taken at or near the height of ridgetop wind project sites and thus may provide a better correlation with the target tower than relatively sheltered surface weather stations. The lowest mandatory monitoring levels

Figure 12-7. A weather balloon used for making soundings in the atmosphere. *Source:* National Oceanic and Atmospheric Administration (NOAA).

are at 1000, 925, 850, and 700 mb, which span the vertical profile from near sea level to roughly 3000 m. In addition, there are several fixed altitudes where wind speed and direction data are required to be reported (8).

Rawinsondes also have disadvantages. There are far fewer rawinsonde stations than surface stations, meaning the nearest rawinsonde station is often much farther away from the project site than the nearest surface station. In addition, the balloons are launched only twice a day, at 12 PM and at midnight Universal Coordinated Time, or UTC (also known as *Greenwich Mean Time*); these days, some weather balloons are launched four times a day, but this practice is relatively recent, and thus, the two extra observations must usually be discarded to ensure a homogeneous long-term data set. With so few observations, a correlation based on daily mean speeds may yield a poor result, and therefore, weekly or monthly means may be called for. This substantially reduces the amount of independent information available for establishing the target–reference relationship.

Despite these disadvantages, it is often a good idea, especially for projects in complex terrain, to obtain rawinsonde data from nearby stations either for direct use in MCP or for verifying the homogeneity through time of available surface weather data.

12.3.4 Modeled Data Sets

In recent years, the use of reference data sets created by atmospheric models has become more common, although it is not yet the industry norm. They are sometimes called *virtual meteorological masts*, or *VMMs*. The most well-known type of modeled data is called *reanalysis data*. It comes in several varieties and is produced by a number of national weather agencies, including the National Centers for Environmental Prediction (NCEP)/National Center for Atmospheric Research (NCAR) and the European Center for Medium-Range Weather Forecasts (ECMWF). The NCEP/NCAR data are available for free and therefore tend to be the most widely used.

All reanalysis data sets are created by using historical weather observations (generally from surface, rawinsonde, satellite, and aircraft-borne instruments) to drive a global or regional NWP model. From these model runs, weather parameters (including temperature, pressure, wind, precipitation) are extracted for every grid point and every level in the model. Reanalysis data sets were created to support climate studies. Unlike real-time weather forecasting models, which are frequently modified, the reanalysis models are fixed for the entire historical simulation.

Reanalysis data have a number of positive attributes, including convenience, multiple levels and types of weather parameters, and a long data record (more than 60 years for some data sets). Because the gridded data are available everywhere covered by the model, there is no difficulty finding suitable grid points. This eliminates much work searching for surface weather stations and data sets, and it provides a common data source for all MCP studies. In parts of the world where surface weather observations are unreliable, reanalysis data (and other modeled data sets) may be the only feasible source of reference data for MCP.

However, reanalysis data also have significant disadvantages and must be used with caution. First, the correlation of the reanalysis winds with tower observations depends on the complexity of the terrain and the resolution of the reanalysis model. The NCEP/NCAR global reanalysis data set, in particular, is relatively coarse, with a resolution of about $2°$ in latitude and longitude (a little over 200 km) and thus may give poor results in mountainous terrain, at coastal boundaries, and in other places where there is a sharp wind gradient.

More importantly, the homogeneity of reanalysis data is limited by that of the observational system used to drive the model, which has changed dramatically over the decades. The bulk of the weather observations in the 1950s and 1960s came from weather balloons supplemented by land and ship-based surface observations. Weather satellites became increasingly important in the 1970s and 1980s, decades that were marked also by the retirement of weather ships, growth in the use of commercial aircraft to supplement weather observations, and a large increase in the frequency of weather observations from both surface and rawinsonde stations (9).

To some degree, the atmospheric model should be able to attenuate the impact of such changes, as observations from one new platform or sensor are reconciled with those available from existing sensors. At different times and in different regions, however, the availability of new data can significantly alter the model's analysis, resulting in spurious trends and shifts in wind and other parameters (10).

In response to these concerns, the concept of a "controlled reanalysis" has been introduced. This approach is similar to reanalysis except that additional care is taken to employ data from a consistent set of observational systems and platforms (such as a fixed number of levels from a fixed set of rawinsonde stations). Research suggests that this method can reduce inconsistencies in traditional reanalysis (11).

In sum, modeled data sets can be useful compliments to surface and rawinsonde observations, but the resource analyst should be wary of relying on them entirely for MCP except when direct observations are unavailable or inadequate to the task. As always, the consistency of the modeled data should be verified through comparisons with independent data sources.

12.4   THE TARGET–REFERENCE RELATIONSHIP

Once the reference station (or stations) is selected, the next step is to establish a relationship between the reference and target winds. This relationship is used to predict the long-term wind resource at the target site based on the entire valid (homogeneous) record of the reference station.

Many types of functional relationships can be used, too many to be described comprehensively here. (Summaries of various methods can be found in References 12 and 13.) The most popular approaches are based on a linear transformation between the reference and target wind speeds (and, occasionally, directions). The general form of the linear equation is the familiar $y = mx + b$, where x is the reference wind speed, y is the target wind speed, m is the slope, and b is the intercept. If the "true" long-term mean wind speed at the reference station is known, then the predicted mean is given by the equation

$$\bar{y} = m\bar{x} + b \qquad (12.4)$$

where the bar over the variable indicates an average. Usually, this equation is determined through a least squares fitting procedure called *linear regression* (see below).

A variety of nonlinear methods (e.g., artificial neural networks or support vector machines) have also been proposed and studied, but they tend to be more complicated than linear methods and require more expertise. Only the linear methods are discussed here.

12.4.1 Data Binning

One distinction between different linear methods is how the data are binned, or grouped in subsets. The so-called bulk method derives a single linear equation for all the data at once. Directional methods, which are also quite popular, construct a different linear equation for each of the several direction bins. Matrix methods go further and bin the data by both direction and speed (and sometimes forego the full linear equation in favor of a ratio) (14). Still other approaches bin the data by time of day (irrelevant when daily averages are used) or time of year.

The bulk method is the simplest to use and probably the most robust, meaning the least susceptible to large errors in inexperienced hands or under far-from-ideal conditions. Other methods require more time and experience. One complication of highly binned approaches is the need to deal with bins that have an insufficient number of counts to provide a reliable fit or ratio. Adjacent bins can be merged or a flexible bin size can be employed to overcome this difficulty.

Whether any particular approach produces a consistently more accurate estimate of the long-term mean wind resource is unclear and depends at least in part on how the objective is defined. When it comes to a single parameter, the mean wind speed, one study employing eight different pairs of reference/target data sets found little difference between three linear methods: a bulk linear regression method, a matrix ratio method, and a variance ratio method (described later).[2] Another study found a very slight reduction in error when employing various directional and time-of-day binning approaches compared to a bulk linear regression approach (15).

When it comes to predicting the wind speed frequency distribution, however, the bulk linear regression method does not perform as well. This issue is addressed below.

12.4.2 Fitting Methods

The simplest method of relating wind speed data from two towers is by taking the ratio of their means (this is effectively a linear equation with $b = 0$). The key problem with this approach is that it assumes a perfect correlation: increasing the reference wind speed by 10% produces a 10% increase in the target speed. If the correlation is actually much less than one, the result can be a substantial error in the predicted long-term mean wind speed, as too much weight is attached to the reference.

Binning approaches that employ ratios can get around this problem to some degree by allowing additional freedom in defining the target–reference relationship. This assumes that the bins span a wide range of wind speeds; directional binning alone may not be enough. A matrix method that bins by speed meets this test.

Alternatively, a linear equation can be established through linear regression. All spreadsheet programs and some commercial wind resource analysis software contain routines to do this. The main thing to know about linear regressions is that they seek to minimize the sum of *squared* errors, meaning they are quite sensitive to large

---

[2] Table 4 in Rogers (2005). A fourth method, which creates a separate linear equation for each component of the horizontal wind vector, performed poorly. Since this approach conflates wind direction and speed, it is not treated as a linear method here.

deviations between prediction and reality. As a result, just a few outlying data points, such as may occur in wind data that have not been properly quality-controlled, can pull the fitted line significantly to one side. For this reason, professional statisticians often prefer other, more robust fitting methods, but for ordinary users, the simplicity and ease-of-use of the linear regression usually make it the method of choice.

Linear regression does not assume a perfect correlation. When the correlation is weak, the slope tends to be small, so variations in the reference wind speed have little effect on the predicted target speed. Another advantage of linear regression is that it can easily incorporate more than one reference station at a time (a multiple linear regression). Sometimes different reference stations capture different aspects of the target site's wind climate; for example, a coastal station may be more representative of the target site than an inland station when the winds come from over the ocean, while the reverse is true when the winds originate from over land. The weight given to each reference station in the fit depends on that station's correlation with the target site and its statistical independence from other stations. A multiple linear regression can be a handy way of improving the overall correlation and allowing an objective determination of the relative value of different stations. However, if too many reference stations are used in multiple linear regression, and especially if they are strongly correlated, there is a risk that the regression will be overspecified. This can produce poor results.

12.4.3 Predicting the Speed Frequency Distribution

A significant drawback of linear regression is that it tends to understate the degree of variation of the target wind speeds, especially when the correlation is weak. For a given linear equation $y = mx + b$, the variance, or standard deviation squared, of the predicted speeds is given by

$$\sigma_y^2 = m^2 \sigma_x^2 \qquad (12.5)$$

The weaker the correlation, the smaller the slope m and therefore the smaller the variance of the predicted wind speeds. To accurately predict the speed frequency distribution, in general, something other than a linear regression is required.

One simple but often effective approach is to scale the observed target site's wind speeds to the predicted long-term mean. Each speed value in the target data set is multiplied by the ratio of the predicted long-term mean to the observed mean:

$$v_i^{(\text{pred})} = \left(\frac{\overline{v}^{(\text{pred})}}{\overline{v}^{(\text{obs})}} \right) v_i^{(\text{obs})} \qquad (12.6)$$

This assumes, in effect, that the data measured at the site accurately capture the relative variation of the wind, so only the mean needs to be adjusted. In practice, this is usually a good assumption: rarely does the estimated power production vary by more than 1–2% because of variations in the wind speed distribution at the same tower, from one year to another, for the same mean.

However, the method only works well if there is at least 1 year of on-site observations. With less than a year of measurements, the seasonal dependence of the speed distribution becomes a concern (as does the accuracy of the predicted mean speed).

The variance ratio method is another way of preserving the target site's variance. The idea is that the slope and intercept of the linear equation $y = mx + b$ are chosen to reproduce the observed variance and mean

$$ y = \frac{\sigma_y}{\sigma_x}x + \left[\overline{y} - \left(\frac{\sigma_y}{\sigma_x}\right)\overline{x}\right] \tag{12.7}$$

The mean values (overbars) are from the concurrent target and reference data sets. The predicted long-term mean speed can be derived from this equation or from the linear regression method. The latter is preferred since only linear regression considers the correlation in determining the size of the MCP adjustment; otherwise, the variance ratio method effectively assumes perfect correlation.

Although the variance ratio method will match the observed speed variance, there is no guarantee that it will produce the correct speed frequency distribution in detail since the relationship between the target and reference speeds may (and usually does) vary with speed, direction, time of day, and other factors. Matrix ratio methods can overcome this particular difficulty with appropriate binning; but even these methods break down when the correlation within each speed and direction bin is not very strong, that is, when there is not a one-to-one correspondence between a particular reference speed and direction and the corresponding target speed and direction. This last shortcoming can be addressed by introducing random noise terms when reconstructing the target data set, but this can only provide an approximate solution.

12.4.4 Direction and Other Parameters

It is usually not necessary to use MCP to predict the target directional distribution, so long as there is at least a year of directional data from the target site. Where the on-site observations are inadequate, the simplest solution is to find the mean offset between the concurrent reference and target directions for each reference direction sector and apply that offset to the full reference data record. This works well, as one might expect, when the directions are highly correlated, but it can break down in less ideal situations. A more general solution is to sample the directional distribution at the target for each reference direction. This method can readily be combined with the matrix sampling method described at the end of the previous section.

Other parameters, such as the observed temperature, can be adjusted to the historical norm using a linear regression between the reference and target site in the same manner as wind speed. If available, air pressure measurements can also be adjusted using this method. The results can be used to adjust the estimated air density at the site.

12.4.5 Summary

While every method has its strong and weak points, it is generally best for the inexperienced analyst to stick with relatively simple, tried-and-true approaches. By this standard, it is hard to beat a bulk linear regression. It is easy to apply, and its estimates of the long-term mean wind speed are about as accurate as any linear method. It is not suitable for predicting the speed distribution; for that, scaling the observed wind speeds to the predicted mean is recommended so long as there is at least 9 months of valid on-site data. More complicated methods are best left to analysts with time and experience, who can fully understand the potential benefits and pitfalls of applying them to the available data sample.

12.5   QUESTIONS FOR DISCUSSION

1. Suppose you observe a significant declining trend in a reference station's data record over a 15-year period, and you are confident of the consistency and quality of the measurements. What are some possible causes of this trend? Given each cause you identify, what are the implications of using this station for MCP?

2. Discuss the potential risks of using a site data record of less than 1 year in MCP.

3. You have a 5-year reference mast at your project site and two masts with 1 year of data at each. When correlating the short-term masts to the reference mast, one yields $r^2 = 0.8$, while the other yields $r^2 = 0.5$. Discuss the potential causes of this difference. If the problem cannot be remedied, what are the uncertainties in the projected 5-year mean for each mast? Assume an interannual variability σ of 4%.

4. What are the relative advantages and disadvantages of surface and upper-air data for MCP? Are there certain project conditions that would cause you to favor one over the other? Are there certain cases where modeled data may be preferable to any source of measured data?

5. Besides distance, what are some other factors that could influence the strength of the wind speed relationship between a target site and a long-term reference station?

6. The majority of surface observation stations measure the wind speed at 10 m height, while wind energy assessments are primarily conducted at much greater heights. Discuss the potential effects of this on an MCP analysis and how those effects can be mitigated.

7. You have 1 year of data from your site, for which the annualized mean wind speed is 6.51 m/s. Your MCP analysis (using linear regression) yields the long-term (LT) mean speeds shown in the table at the top of the following page for 10 different prospective reference stations. Which stations would you include in your analysis, and why? Can you think of two methods of arriving at your final long-term mean speed estimate?

8. You are considering a multiple linear regression with two reference stations. Station A's record starts in 1995 and station B's in 1980. Do you think it is reasonable to go back to 1980? Why or why not? Describe the process you might use to decide which years to use as the start of both station A's and

Reference station	Slope	Intercept	r^2	Projected LT speed
A	1.24	2.15	0.90	6.39
B	0.88	0.45	0.89	6.56
C	0.90	0.54	0.89	6.59
D	1.36	2.04	0.89	6.25
E	1.30	1.44	0.88	6.31
F	1.30	0.82	0.83	6.24
G	1.26	1.39	0.76	6.36
H	1.19	1.89	0.62	6.37
I	0.38	2.49	0.61	6.35
J	1.06	2.85	0.47	6.68

station B's reference periods. Then outline the steps and necessary equations to determine the long-term hub-height speed at your on-site mast.

9. Refer to the trend plot presented below. Suppose all five stations have a good correlation with your site. For each station labeled A–E, comment on its suitability for long-term climate adjustments and whether or not you would consider using it.

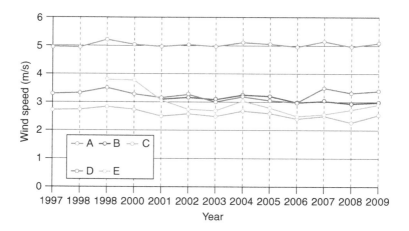

10. Discuss the advantages and disadvantages of at least three out of the following five methods of relating target and reference data sets: (i) ratio of mean speeds; (ii) linear regression constrained through the origin; (iii) unconstrained linear regression; (iv) matrix method with binning by direction and speed; and (v) artificial neural network.

11. Why is it important to avoid reducing the variability of wind speeds when creating a time series of target data reflecting long-term historical conditions?

Which methods discussed in this chapter accomplish this? How would reducing the variability impact the estimated energy production, in general?

REFERENCES

1. Vautard R, Cattiaux J, Yiou P, Thépaut J-N, Ciais P. Northern Hemisphere atmospheric stilling partly attributed to an increase in surface roughness. Nat GeoSci 2010;3:756–761.
2. Pryor SC, Barthelmie RJ, Young DT, Takle ES, Arritt RW, Flory D, Gutowski Jr. WJ, Nunes A, Roads J. Wind speed trends over the contiguous United States. J Geophys Res 2009;114:D14105. doi: 10.1029/2008JD011416, 2009.
3. Freedman JM, Zack JW. Climate and Wind Energy: Outlook for Wind in a World Undergoing Climate Change—Recent Trends in Wind Speed, American Wind Energy Association Wind Power 2007, June 2007.
4. Breslow PB, Sailor DJ. Vulnerability of wind power resources to climate change in the continental United States. Renew Energy 2002;27:585–598.
5. Sailor DJ, Smith M, Hart M. Climate change implications for wind power resources in the Northwest United States. Renew Energy 2008;33:2393–2406.
6. Freedman JM. Final Report: effects of climate change on production of electricity from wind power in California. California Energy Commission; 2009.
7. Ren D. Effects of global warming on wind energy availability. J Renew Sustain Energy 2010;2:052301.
8. Federal Meteorological Handbook No.3. Rawinsonde and Pibal Observations, Office of the Federal Coordinator for Meteorology, 29 May 2007.
9. Kistler R, Kalnay E, Collins W, Saha S, White G, Woollen J, Chelliah M, Ebisuzaki W, Kanamitsu M, Kousky V, van den Dool H, Jenne R, Fiorino M. The NCEP/NCAR reanalysis. Bull Am Met Soc 2001;8:247–267.
10. Brower MC. The Use of NCEP/NCAR Reanalysis Data in MCP, AWS Truepower; 2006.
11. Waight K. Validation of MASS wind anomaly simulations: 1997–2006, AWS Truepower; March 2008.
12. Rogers A, Rogers J, Manwell J. Comparison of the performance of four measure-correlate-predict algorithms. J Wind Energy Ind Aerodyn 2005;93:243–264.
13. Thoegersen ML. Measure-correlate-predict methods: case studies and software implementation. Milan, Italy: European Wind Energy Conference; 2007.
14. Anderson M. A Review of MCP techniques. UK: Renewable Energy Systems Ltd.; 2004.
15. Oliver A, Zarling K. Time of day correlations for improved wind speed predictions. Chicago, Illinois: American Wind Energy Association, Windpower 2009, May 2009. Renewable Energy Systems, Ltd.

SUGGESTIONS FOR FURTHER READING

Rogers A, Rogers J, Manwell J. Comparison of the performance of four measure-correlate-predict algorithms. J Wind Energy Ind Aerodyn 2005;93:243–264. Available at http://www-unix.ecs.umass.edu/~arogers/_html_version/publications/_publicationpdfs/AnthonyRogers_2005_JWEIA_Measure_Correlate_Predict.pdf. (Accessed 2012).

Sarachik S, Crane M. The El Ni no-Southern oscillation phenomenon. UK: Cambridge University Press; 2010.

Solomon S, Qin D, Manning M, Chen Z, Marquis M, Averyt KB, Tignor M, Miller HL, editors. Contribution of Working Group I to the Fourth Assessment Report of the Intergovernmental Panel on Climate Change, 2007. UK: Cambridge University Press; 2007. Available at http://www.ipcc.ch/publications_and_data/ar4/wg1/en/contents.html. (Accessed 2012).

13

WIND FLOW MODELING

The main purpose of wind flow modeling is to estimate the wind resource at every proposed or potential wind turbine location so that the wind plant's energy production can be calculated and its design optimized. This usually means extrapolating from the wind resource measured at one or more meteorological towers using a numerical wind flow model of some kind.

In an ideal world, wind flow modeling, just like shear and long-term climate adjustments, would not be necessary. Wind measurements would be taken at every likely turbine location to eliminate any possibility of significant error. However, for most projects, this would be an expensive proposition. In practice, wind flow modeling is an essential part of the wind resource practitioner's toolkit. It is also one of the largest sources of uncertainty in most energy production estimates.

Aside from estimating the variation in the wind resource across the project area, wind flow modeling must account for each turbine's influence on the operation of other turbines, the so-called wake effect. Wake modeling is usually performed separately from wind flow modeling using specialized software. It is discussed in Chapter 16.

Unlike most other chapters in this book, this chapter does not present a preferred method or modeling approach. There are simply too many methods with very

Wind Resource Assessment: A Practical Guide to Developing a Wind Project, First Edition.
Michael Brower et al.
© 2012 John Wiley & Sons, Inc. Published 2012 by John Wiley & Sons, Inc.

diverse characteristics and applications to take such a position. Instead, we provide an overview of the different modeling approaches that are available, including their strengths and weaknesses, and establish some general guidelines applicable to all methods—and most important, the appropriate use of measurements to manage and limit errors.

13.1 TYPES OF WIND FLOW MODELS

Spatial modeling approaches can be conveniently classed in four general categories: conceptual, experimental, statistical, and numerical.

13.1.1 Conceptual Models

Conceptual models are theories describing how the wind resource is likely to vary across the terrain. They are usually based on a combination of practical experience and a theoretical understanding of boundary layer meteorology.

A very simple conceptual model might state that the wind resource at one location (a turbine) is the same as that measured at a different location (a met mast). This could be quite a good model in relatively flat terrain or along a fairly uniform ridgeline, for example. Where the terrain and land cover vary substantially, a more nuanced picture is usually required. This might include theories concerning the influence of elevation on the mean wind speed, the relationship between upwind and downwind slope and topographic acceleration, channeling through a mountain gap, and the impact of trees and other vegetation. These concepts or theories are then turned into practical recommendations for the placement of wind turbines, accompanied by estimates of the wind resource they are likely to experience.

As wind projects become larger and are built in ever more varied wind climates, it becomes more and more difficult to implement a purely conceptual approach in a rigorous or repeatable way. Nevertheless, a good conceptual understanding of the wind resource is an invaluable asset in all spatial modeling. Most important, it provides a check on the reasonableness of other methods. A good conceptual understanding is better than a bad numerical model or a good numerical model that is wrongly applied.

13.1.2 Experimental Models

Experimental in this context refers to creating a sculpted scale model of a wind project area (such as that shown in Figure 13-1) and testing it in a wind tunnel. (This is also known as *physical modeling*, a term we avoid because of possible confusion with numerical wind flow models, which are based on physical principles.) The conditions in the wind tunnel, such as the speed and turbulence, must be matched to the scale of the model to replicate real conditions as closely as possible. While the wind tunnel is running, the wind speeds are measured at various points on the scale model using tiny anemometers (usually hot-wire anemometers). The results form a picture of how

Figure 13-1. Scale model of the Altamont Pass used in wind tunnel tests. *Source:* Lubitz WD, White RB. Prediction of wind power production using wind-tunnel data, a component of a wind power forecasting system. Proceedings of AWEA WindPower 2004.

the wind varies across the site. The relative speeds between points are then usually related to a mast where the speeds have been measured in the field.

Although studies comparing experimental methods to other methods are scarce, there is no reason to think this type of approach cannot work well under many conditions. It may even provide unique insights in areas where numerical wind flow models are prone to fail, such as near the edge of a steep cliff. However, few wind resource analysts adopt this method because of the time and special skills required to build an appropriate model and the need for access to a wind tunnel. In addition, the method has some limitations (such as the difficulty of modeling thermally stable conditions and the challenge of appropriately matching atmospheric parameters to the physical scale).

13.1.3 Statistical Models

Statistical models are based on relationships derived entirely or primarily from on-site wind measurements. Typically, one tests different predictive parameters, such as elevation, slope, exposure, surface roughness, and other indicators, to find those that appear to have the strongest relationship with the observed wind resource at several masts. In principle, any parameters can be used, although in practice it makes sense to focus on those for which there is a reasonable theoretical basis for believing a relationship should exist. This is one place where a good conceptual understanding is helpful.

It is probably easiest to explain this approach by example. Suppose one has measured the mean wind speeds at several different towers at different points within a wind resource area. Suppose the speeds are plotted against, say, elevation, and a strong correlation is found. From this relationship, a linear equation ($y = mx + b$) could be derived and then applied to predict the speed at any other point in the area.

Statistical models are appealing because they are well grounded in measurement and are fairly simple and transparent, unlike numerical wind flow models, which often seem like "black boxes." And they can work surprisingly well, particularly for wind climates driven by synoptic-scale winds (i.e., where thermally driven mesoscale circulations are largely absent), which tend to exhibit the clearest relationships between wind speed and certain topographic indicators such as elevation and exposure. Figure 13-2 illustrates the relationship observed at 74 towers in seven wind resource areas between variations in wind speed and downwind exposure, defined as the difference between the elevation of a given point and the average elevation out to 3000 m in the downwind direction.

One of the potential limitations of statistical methods is that they can produce larger-than-expected errors when making predictions outside the range of conditions used to train the model. Suppose, for example, that one has data from three towers at varying elevations along a ridgetop. Will the relationship between mean speed and, say, elevation implied by these three towers hold when predicting the speed off the ridgetop? Not necessarily, because the topographic influence on the wind flow may be very different at the top compared to that at the slopes. In this respect, statistical models can be less reliable than numerical wind flow models, which are designed to produce plausible (if not accurate) results in a wide range of conditions.

Determining the accuracy of a statistical model is a particular challenge of this approach. To derive an objective estimate of the uncertainty, it is necessary to divide the data set into two groups: one to train the model and the other to validate the model.

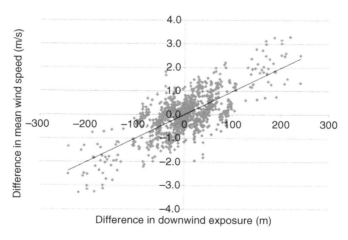

Figure 13-2. Data from pairs of 74 towers in seven wind resource areas indicate a significant relationship between the differences in mean speed and downwind exposure. Such a statistical relationship can be used to predict variations in wind speed across a project area. *Source:* AWS Truepower.

The most rigorous procedure is to derive empirical relationships by fitting variables and functions to the "training" data and then determine the error with respect to the "validation" data that have been withheld. Many sites lack sufficient data with which to conduct such a validation. In such cases where the validation data must be included in the training data set, there is a tendency to underestimate the errors.

Nevertheless, statistical models are a valid approach when proper procedures are followed. Statistical methods can also be combined with other approaches, such as numerical wind flow models. A good example of this is the ruggedness index (RIX) correction that is sometimes used with the Wind Atlas Statistical Package (WAsP) model (described below). RIX is a parameter that has been found through statistical modeling to be a good predictor of WAsP errors in some circumstances (1).

13.1.4 Numerical Wind Flow Models

The most popular methods of spatial modeling rely mainly on numerical wind flow models. There are numerous wind flow models in use by the wind industry today, which are based on a variety of theoretical approaches. All models attempt to solve at least some of the physical equations governing motions of the atmosphere, with varying degrees of complexity. The models fall into four general categories: mass-consistent, Jackson–Hunt, computational fluid dynamics (CFD), and mesoscale NWP models.

Mass-Consistent Models. The first generation of wind flow models developed in the 1970s and 1980s (e.g., NOABL (2), MINERVE) were mass-consistent models, so called because they solve just one of the physical equations of motion, that governing mass conservation. When applied to the atmosphere (assuming it is incompressible, a good assumption within the boundary layer), the principle of mass conservation implies that wind forced over higher terrain must accelerate so that the same volume of air passes through the region in a given time. As a result, these models predict stronger winds on hilltops and ridgetops and weaker winds in valleys. They cannot handle thermally driven wind patterns, such as sea breezes and mountain-valley circulations, and flow separations on the lee side of hills or mountains.

The solution offered by a mass-conserving model is not unique: the governing equation actually permits an infinite number of solutions. Instead, most models are designed to depart by the smallest possible amount from an initial wind field "guess" derived from observations or another model (e.g., a NWP model run at a coarser resolution). Such a characteristic sets this type of model somewhat apart from other numerical models, which make no such assumption. It also means that mass-consistent models are able to take advantage of data from additional meteorological towers in a natural way, by modifying the initial guess.

Jackson–Hunt Models. The next generation of models (e.g., WAsP (3, 4), MS-Micro or MS3DJH (5, 6), Raptor (7), Raptor NL (8)) were originally developed in the 1980s and 1990s based on a theory advanced by Jackson and Hunt (9). They go beyond mass conservation to include momentum conservation by solving a linearized form of the Navier–Stokes equations governing fluid flow. The most important simplification in the Jackson–Hunt theory is that the terrain causes a small perturbation to an otherwise constant background wind. This assumption allows the equations to be solved using a fast numerical technique.

No spatial modeling chapter would be complete without a discussion of WAsP (Fig. 13-3), a Jackson–Hunt model developed by the Risø National Laboratory of Denmark, which has been and probably remains the most widely used numerical wind flow model in the wind industry. The "WAsP method" (Fig. 13-4) is deeply entrenched in spatial modeling practice, especially in Europe. It proceeds in two stages. First, the observed wind at a mast is used to derive the background wind field, which represents

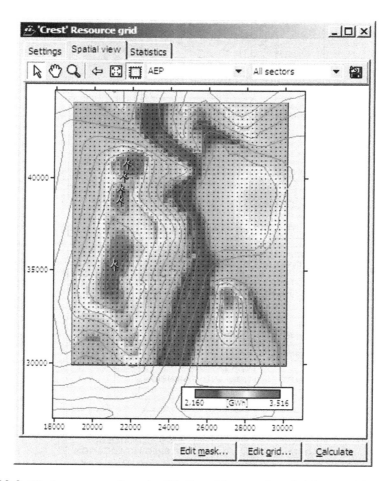

Figure 13-3. Wind power map from the WAsP model, a popular wind flow modeling application. Like other Jackson–Hunt models as well as mass-consistent models, WAsP captures the tendency of wind speed to increase over high ground and decrease in valleys. Ridges oriented perpendicular to the flow exhibit the greatest topographic acceleration. *Source:* Risø National Laboratory.

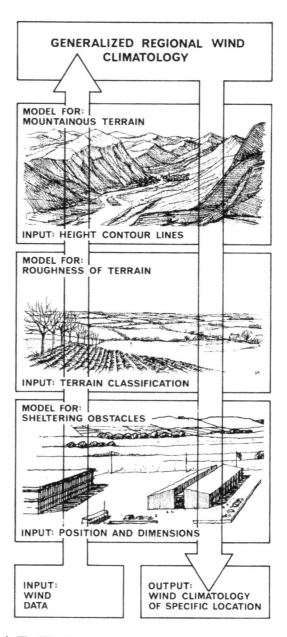

Figure 13-4. The WAsP mapping process. *Source:* Risø National Laboratory.

the wind resource that would exist in the absence of terrain. This background wind field is typically summarized in a file known as a *wind atlas* or *library file*. Second, the process is subsequently reversed using the background wind as an input to predict the wind profile at other points.

In addition to implementing the basic Jackson–Hunt approach, WAsP contains several modules that address various needs in wind flow modeling, including the ability to incorporate the effects of surface roughness changes and obstacles. Perhaps, because WAsP was developed in the relatively flat terrain of northern Europe where roughness changes and obstacles are among the main factors influencing the wind resource, these modules have been developed to quite an advanced degree.

It is widely recognized that WAsP, along with other Jackson–Hunt models, is not equipped to handle complex terrain. ("Complex" in this context is usually defined as terrain where the slope exceeds 30% over a significant portion of the area.) The essential problem is that steep terrain induces changes in the wind flow that do not satisfy the assumption of a small perturbation. The overlooked effects may include recirculation behind cliffs, flow separations at abrupt changes in slope, and vertical winds.

The WAsP model also ignores effects of thermal stability and temperature gradients. Thermal stratification and buoyancy forces can have a large influence on the response of wind to terrain. When the boundary layer is thermally stable, the air near the surface is cooler and denser than the air aloft. The wind, therefore, resists going over higher terrain and instead seeks a path around it, through channels or gaps, or is blocked. Indeed, when WAsP was first applied in coastal mountain passes where the first US wind farms were built, it failed to predict the wind resource distribution accurately. This experience gave wind flow modeling something of a bad name in US wind resource assessment circles for many years.

Despite its known limitations, WAsP remains very popular. This is partly because many wind project sites do not involve very steep terrain or significant mesoscale circulations. In addition, practical steps can be taken to improve the results. One method is to install additional masts so as to limit the distance over which the model must extrapolate the resource. Also, in some instances, it has been shown that errors can be reduced through the RIX adjustment. The RIX parameter represents the proportion of terrain upwind of a point exceeding a certain slope threshold, such as 30%. Relative wind speeds between two points are adjusted according to a simple formula that depends on the difference in RIX between them. Experiments have shown that this adjustment can be quite effective.

Finally, more sophisticated models (described below) are sometimes no more accurate than the first- or second-generation wind flow models. This last point reflects a hard truth of atmospheric modeling: it is exceptionally difficult to do it well, and sometimes it is better to ignore aspects of wind flow one cannot simulate well than to simulate them poorly.

CFD Models. As personal computers have grown more powerful, it is natural that CFD models, computer programs designed originally to model turbulent fluid flows for airplane bodies, jet engines, and the like, would be turned to the task of spatial wind resource modeling.

The critical difference between CFD and Jackson–Hunt models is that CFD models solve a more complete form of the equations of motion known as the *Reynolds-averaged Navier–Stokes* (RANS) *equations*. They do not assume the terrain induces a small perturbation on a constant wind field. This means they are capable of simulating nonlinear responses of the wind to steep terrain, such as flow separation and recirculation (Fig. 13-5). They also do not have to make certain other simplifying assumptions, such as that shear stress and turbulence act only near the surface. This, in turn, allows CFD models to simulate the influences of roughness changes and obstacles directly. (WAsP and other linear flow models, in contrast, generally do this in separate modules.)

CFD models represent an important new tool in the resource analyst's toolkit. Among their advantages are that they provide an independent picture of the wind resource that often looks quite different from that generated by Jackson–Hunt models such as WAsP, and they can provide useful information concerning turbulence intensities, shear, direction shifts, and other features of wind flow in complex terrain.

CFD models have sometimes shown very good agreement to wind tunnel experiments for 2D and 3D flows around idealized escarpments and steep hills, even on the lee side with the recirculation zone (10, 11). Some tests using field data have likewise demonstrated significant improvements over linear models, e.g., on the order of a 50% reduction in root-mean-square error in a sample of 14 mast pairs (12). Other side-by-side tests have reached a different conclusion. One experiment at 4 wind project sites with a range of terrain conditions found that a leading CFD model produced a 22% larger error, on average, than WASP (18). The Bolund experiment undertaken by the Risø National Laboratory (Denmark) involving more than 35 different CFD models showed that "the average overall error in predicted mean velocity of the top ten models (all RANS-based) was on the order of 13–17% for principal wind directions" (13)—a disappointing result. It seems much depends on the particular situation, CFD model, and (perhaps) the user's expertise.

Wind Sim

Figure 13-5. CFD models such as WindSim, depicted here, are capable of simulating nonlinear flow features as recirculation behind steep terrain. *Source:* WindSim AS.

Clearly, the success of CFD modeling is not assured, and there is a continuing need to validate CFD results with high quality wind measurements. Problems have been ascribed to various factors, including inaccuracies in initial and boundary conditions (which are usually assumed to be homogeneous and follow a neutrally stratified, logarithmic profile), limited grid resolution, and treatment of turbulence. The added complexity of the models may be a problem, as some users may not be equipped to run them properly. Another factor is that in general, CFD models are not designed to take into account any circulations due to temperature gradients. The lack of a complete prognostic equation for temperature in CFD models is, in turn, the result of another assumption made in most CFD models, which is that the wind flow is steady state. In a manner not unlike WAsP, most CFD models assume a constant incoming wind field.

Mesoscale Numerical Weather Prediction Models. The last class of wind flow models covered in this chapter is the mesoscale NWP model. This type of model has been developed primarily for weather forecasting. Like CFD models, mesoscale models solve the Navier–Stokes equations. Unlike CFD models, however, they include parameterization schemes for solar and infrared radiation, cloud microphysics and convection (cumulus clouds), a soil model, and more. Thus, they incorporate the dimensions of both energy and time and are capable of simulating such phenomena as thermally driven mesoscale circulations (such as sea breezes) and atmospheric stability, or buoyancy. In the world of mesoscale modeling, as in the real world, the wind is never in equilibrium with the terrain because of the constant flow of energy into and out of the region, through solar radiation, radiative cooling, evaporation, and precipitation, a cascade of turbulent kinetic energy down to the smallest scales and dissipation into heat.

Mesoscale models consequently offer considerable hope for simulating wind flows accurately in complex terrain. They have, however, one big drawback: they require enormous computing power to run at the scales required for the assessment of wind projects. The typical model resolution for most mesoscale simulations is on the order of kilometers, meaning a single grid cell is kilometers across. It is clearly impossible to obtain a detailed picture of the wind resource within a project area at such a scale.

One way around this problem is to couple mesoscale models with a microscale model of some kind. This could be a statistical model, if there is sufficient on-site wind data to create reliable statistical relationships. More often, it is a simplified wind flow model, usually either a mass-consistent model or a Jackson–Hunt model. Examples include AWS Truepower's MesoMap and SiteWind systems (14), 3TIER's FullView system, the Risø National Laboratory's KAMM–WAsP system (15), and Environment Canada's AnemoScope system (16).

Research suggests, not surprisingly, that such methods can be more accurate than simplified wind flow models where mesoscale effects play a significant role. One example is the wind resource in a coastal mountain pass such as the Altamont Pass in California. Here, a model such as WAsP predicts that the best winds should be at the top of the pass, whereas a mesoscale–microscale modeling approach predicts the acceleration of the relatively cool and dense marine air mass as it flows down the slope. The result is a definite improvement in accuracy (Fig. 13-6).

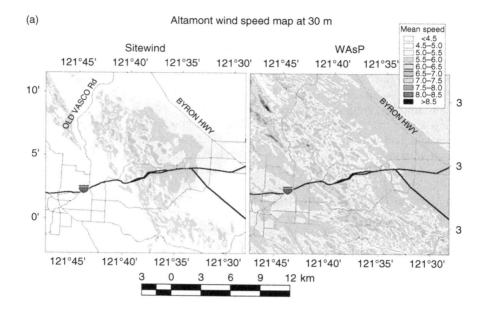

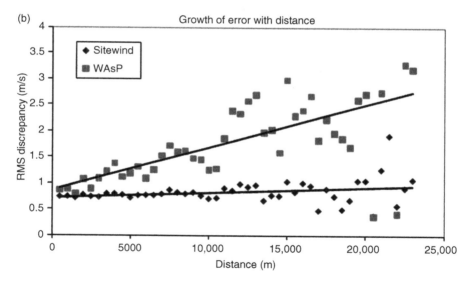

Figure 13-6. (a) Wind resource maps of the Altamont Pass, California, created by the SiteWind mesoscale–microscale modeling system (left) and by WAsP (right). The SiteWind map shows the wind resource concentrated on the eastern slopes, the result of gravity acting on relatively cool, dense marine air. WAsP suggests the wind resource is more widely distributed and is at a maximum at the top of the pass. (b) Comparison with observations indicates the mesoscale–microscale modeling is more correct, and that errors grow more rapidly with distance from the reference mast for WASP than for Sitewind (17).

Altamont Pass represents a severe test for other models because the masts span a large distance, nearly 25 km, and mesoscale circulations are the key factor determining the spatial distribution of the wind resource. In project areas dominated by synoptic-scale winds, on the other hand, the advantages are less decisive, and they are likely to diminish further or disappear at short distances from observational masts. Below the effective scale of the mesoscale model, say, a few hundred meters to a few kilometers, the microscale model dominates, and no improvement over conventional modeling is possible.

Ultimately, it would be desirable to adapt a mesoscale model to run directly at the resolution required for wind plant micrositing (e.g., 50 m), thus eliminating the need for a simplified downscaling approach. In this mode, and with appropriate modifications, the models begin to operate like large-eddy simulation models, which are used to simulate nonsteady flows at very high resolution (meters). This approach is becoming feasible with modern high performance computing clusters. However, the potential improvement in accuracy relative to the computer cost has not been evaluated. If successful, such an approach will most likely be reserved for the largest and most complex wind project areas where the additional cost can be justified.

13.2 APPLICATION OF NUMERICAL WIND FLOW MODELS

With such a wide variety of options for spatial modeling, it is clear that no single approach can be recommended above all others and in every circumstance. To narrow the discussion a little, we focus on numerical wind flow models, which are by far the most widely used. The following guidelines are applicable to such models.

13.2.1 Topographic Data

Accurate, high resolution topographic data are essential for all numerical wind flow modeling. A typical spatial resolution for modeling is 50 m (often the data are resampled to this scale from a higher resolution). Resolutions as high as 10 or 30 m are sometimes used, but it is unclear that they confer much benefit in overall accuracy since turbine rotors are typically 70–100 m in diameter, and turbines effectively average the wind resource over this area.

In the United States, the preferred source of topographic data is the US Geological Survey (USGS) National Elevation Dataset (NED), with a spatial resolution of 10 or 30 m (except in Alaska, where the resolution is 60 m). Outside the United States, data of similar quality are usually available from national mapping agencies, or absent anything better, analysts may use the Shuttle Radar Topography Mission data set, with a resolution of approximately 90 m.

13.2.2 Land Cover Data

It is likewise important to employ accurate, high resolution land cover data. Most modern land cover data sets are created from satellite imagery, such as those produced by the Landsat and Satellite Pour l'Observation de la Terre (SPOT) satellites. Since the land cover classifications are derived by computer software from spectral measurements and have only been checked in certain areas, they are subject to error. Their accuracy should be verified, where possible, through aerial and ground photographs.

In the United States, the preferred data set is the National Land Cover Database (NLCD 2001), a 30-m resolution data set derived from Landsat Thematic Mapper imagery. A similar data set known as EarthSat GeoCover is available for purchase in other regions of the world from MDA Federal, Inc. Many countries have home-grown land cover data sets, and Europe is covered by the Coordination for Information on the Environment (CORINE) system.

In most cases, the land cover classifications must be converted to roughness values (in meters) to be used in numerical wind flow modeling. There is no universally accepted roughness conversion system. Table 13-1 presents the ranges of values used by AWS Truepower.

For more precise land cover information, some private companies (e.g., Intermap Technologies) offer high resolution surface models that include features such as buildings, vegetation, and roads, in addition to terrain elevation data.

13.2.3 Mast Number and Placement

To achieve the standards of accuracy required for energy production estimates for utility-scale wind projects, all wind flow modeling must be anchored in high quality observations from the project area. A minimum of one mast, equipped and installed to the specifications described in this book, is recommended, although a remote sensing system may serve as the primary observational platform in some cases.

Wind resource data should be collected at locations representing the full range of wind conditions likely to be encountered by the wind turbines in the project. Where this criterion is not met, the uncertainty in the spatial modeling increases substantially. This criterion is sometimes translated into distance, as in the guideline that no turbine should be placed farther than 1–3 km from a met mast in complex terrain (Table 3-2). But distance is only one factor to consider; differences in elevation, topographic exposure, slope, and aspect (angle of slope with respect to north) may also be important.

In general, the greater the number of masts deployed, the smaller the uncertainty in the predicted resource. However, this is true *only* if the masts are well distributed throughout the proposed turbine array. If the masts are unevenly distributed, then the benefit of the multiple masts is reduced. In the worst case, if the masts are clumped in one area while the turbines are in a different area, then the additional masts provide little, if any, benefit. Examples of preferred and poor mast placement are provided in Figure 13-7. Uncertainty is discussed further in Chapter 15.

Table 13-1. Roughness ranges for typical land use/land cover categories. Values may vary with geographic location

Land cover type	Roughness range, m
Water	0.001
Urban/developed area	0.3–0.75
Forest	0.9–1.125
Wetland	0.15–0.66
Shrub land	0.1–0.2
Cropland	0.03–0.07

Source: AWS Truepower.

13.2.4 Adjustments to Multiple Masts

Most numerical wind flow models are equipped to use data from just one mast. These days, however, wind projects usually employ several masts. This raises the question of how to make the best use of the additional data. Since any blending of predictions from different masts is done in most cases in the plant design software rather than in the wind flow software, this topic is addressed in Chapter 16.

13.3 QUESTIONS FOR DISCUSSION

1. Suppose you are developing a conceptual model for a wind project on a ridgetop site. Would your model assume that the wind speeds increase or decrease with

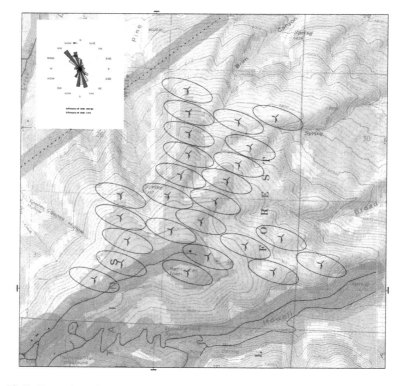

Figure 13-7. Examples of preferred and poor mast placement for a hypothetical turbine layout. The four masts in the lower left-hand corner, indicated by black dots, are poorly situated for this proposed turbine array. The single mast near the center of the array (red dot) is better situated, but since it is on the highest ground, it may not accurately capture the wind resource for some turbines at lower elevations, even with numerical wind flow modeling. *Source:* AWS Truepower.

elevation? Do you think the relationship between speed and elevation should be the same for variations in elevation along the ridgeline compared to down the slope? Explain why or why not.

2. You are developing a conceptual model for a wind project at an inland, flat site. Would your model assume that the wind speeds increase or decrease as the vegetation gets denser and taller? Explain why.

3. Why does the principle of mass conservation imply that wind forced over higher terrain must accelerate?

4. Using the Internet or any other reference at your disposal, identify the principles of conservation that underlie the Navier–Stokes equations. Would you conclude that a model that solves the Navier–Stokes equations provides a more complete depiction of the influence of terrain on wind flow than a mass-conserving model? Why or why not?

5. Both mass-conserving and Jackson–Hunt wind flow models have trouble in complex terrain. Explain some of the reasons why. In this context, to be considered a "complex" site, how steep must the slopes reach over a significant portion of the area?

6. Using the Internet or any other sources at your disposal, find out when the WAsP model was initially developed. Also determine what the typical clock speed of a personal computer was at that time. What is the clock rate of typical personal computer now?

7. What are the main differences between a CFD model and a Jackson–Hunt model? What are the main differences between a mesoscale NWP model and a CFD model? Which of these model types provides the most complete depiction of the various forces affecting the wind?

8. Apart from the three vector components of the wind, what are the other four main meteorological variables that are predicted by a mesoscale NWP model? (Use the Internet if necessary.)

9. What is the typical resolution of a mesoscale NWP model? Why are these models not run at much higher resolution? What is the current modeling approach to get around this problem?

10. Consider three wind project sites, all about 10 km across: (i) a series of gently rolling hills of relative height 100 m and maximum slope 5%, (ii) a steep ridgeline of relative height 500 m and maximum slope 50%, and (iii) a coastal mountain pass. Which general types of model should be able to handle each situation adequately?

REFERENCES

1. Mortensen NG, Bowen AJ, Antoniou I. Improving WASP predictions in (too) complex terrain. In: Proceedings of the European Wind Energy Conference; Athens, Greece; 2006.

2. Phillips GT. A preliminary user's guide for the NOABL objective analysis code. La Jolla (CA): Report from Science Applications, Inc.; 1979.115.

3. Troen I, Petersen EL. European wind atlas. Roskilde, Denmark: Report from the Risoe National Laboratory; 1989.

4. Troen I. A high resolution spectral model for flow in complex terrain. Proceedings from the 9th Symposium on Turbulence and Diffusion, Roskilde, Denmark; 1990.

5. Beljaars ACM, Walmsley JL, Taylor PA. A mixed spectral finite-difference model for neutrally stratified boundary-layer flow over roughness changes and topography. Boundary-Layer Meteorol 1987;38:273–303.

6. Taylor PA, Walmsley JL, Salmon JR. A simple model of neutrally stratified boundary-layer flow over real terrain incorporating wave number-dependent scaling. Boundary-Layer Meteorol 1983;26:169–189.

7. Ayotte KW, Taylor PA. A mixed spectral finite-difference 3D model of neutral planetary boundary-layer flow over topography. J Atmos Sci 1995;52:3523–3537.

8. Ayotte KW. A nonlinear wind flow model for wind energy resource assessment in steep terrain. In: Proceedings of Global WindPower Conference, Paris, France; 2002.

9. Jackson PS, Hunt JCR. Turbulent wind flow over low hill. Q J Roy Meteorol Soc 1975; 101:929–955.

10. Bitsuamlak GT, Stathopoulos T, Bédard C. Numerical evaluation of wind flow over complex terrain: Review. J Aerosp Eng 2004;17:135–145.

11. Murakami S, Mochida A, Kato S. Development of local area wind prediction system for selecting suitable site for windmill. J Wind Eng Ind Aerodyn 2003;91:1759–1775.

12. Pereira R, Guedes R, Santos CS. Comparing WAsP and CFD wind resource estimates for the 'Regular' User. In: Proceedings of the European Wind Energy Conference; Warsaw, Poland. 2010.

13. Sumner J, Sibuet Watters C, Masson C. CFD in wind energy: the virtual, multiscale wind tunnel. Energies 2010;3:989–1013.

14. Brower M. Validation of the WindMap Program and Development of MesoMap. In: Proceeding from AWEA's WindPower conference; Washington (DC); 1999.

15. Frank HP, Rathmann O, Mortensen N, Landberg L. The Numerical Wind Atlas, the KAMM/WAsP Method. Riso-R-1252 report from the Risoe National Laboratory, Roskilde, Denmark; 2001. p. 59.

16. Yu W, Benoit R, Girard C, Glazer A, Lemarquis D, Salmon JR, Pinard J-P. Wind Energy Simulation Toolkit (WEST): a wind mapping system for use by the wind-energy industry. Wind Eng 2006;30:15–33.

17. Reed R, Brower M, Kreiselman J. Comparing SiteWind with standard models for energy output estimation. In: Proceedings from EWEC; London; 2004.

18. Beaucage P, Brower M. Evaluation of four numerical wind flow models for wind resource mapping. Submitted to the related special issue of the Wind Energy Journal (June 2011).

SUGGESTIONS FOR FURTHER READING

Ferziger JH, Peric M. Computational methods for fluid dynamics. USA: Springer-Verlag Ed; 2002.

Kalnay E. Atmospheric modeling, data assimilation and predictability. UK: Cambridge University Press; 2003.

Pielke RA. Mesoscale meteorological modeling. USA: Academic Press; 2002.

Stensrud DJ. Parameterization schemes: keys to understanding numerical weather prediction models. UK: Cambridge University Press; 2007.

Stull RB. An introduction to boundary layer meteorology. USA: Kluwer Academic Publishers; 1988.

Troen I, Petersen EL. European wind atlas. Roskilde, Denmark: Report from the Risoe National Laboratory; 1989.

14

OFFSHORE RESOURCE ASSESSMENT

A growing share of the world's wind energy development is now occurring offshore, in both lakes and oceans. In part, this is because many offshore areas have a good wind resource. Even more important, they offer opportunities for large-scale development that may not otherwise be available near densely populated and protected coastal areas. This point helps explain why most of the world's offshore wind projects have been built or are being planned in Western Europe, where the population density is high and there is relatively little land available for large wind projects. To date, most offshore wind projects have been built less than 25 km from the shore in water of less than 40 m depth. With improvements in turbine foundation technologies (including floating platforms) and installation practices, projects in the future may be sited farther from shore and in deeper waters.

Wherever offshore projects are sited, their success, like that of onshore projects, depends on sound wind resource assessment. Many of the basic principles and guide-lines of resource assessment covered in this book apply equally offshore. For the most part, the instruments and parameters measured are the same, as are the methods used to collect and QC the data, characterize the resource, project the measurements to

Wind Resource Assessment: A Practical Guide to Developing a Wind Project, First Edition.
Michael Brower et al.
© 2012 John Wiley & Sons, Inc. Published 2012 by John Wiley & Sons, Inc.

the turbine hub height, correct for short-term climate variability, and extrapolate the resource to turbine locations using numerical wind flow modeling.

Nevertheless, offshore environments offer unique challenges. These include a scarcity of existing, high quality wind measurements taken at or near turbine hub heights; the cost and delays involved in deploying measurement platforms offshore; the need to account for the effects of currents, waves, and (in cold climates) ice in the design of the platforms; the difficulty of reaching offshore sites for maintenance and repair, especially in rough weather; heightened safety risks to field personnel; equipment exposure to corrosive elements such as sea spray, salt water, and high humidity; and extreme storm-driven winds and waves. These challenges mean that an offshore measurement campaign usually costs much more than an onshore campaign of similar scope.

This chapter focuses on the special considerations and requirements for offshore resource assessment. It covers the nature of the offshore wind environment; the variety of monitoring station types and instrumentation, including remote sensing, that can be used for resource assessment; ancillary systems such as power supplies, data loggers, and communications; the challenges of offshore installation; and operations and maintenance.

Aside from monitoring the wind resource, the design and permitting of offshore wind projects requires characterizing other aspects of the ocean or lake environment, including currents, waves, ice, and water temperature; the geophysical characteristics of the seabed or lakebed; and fish, birds, and other wildlife. Although not a focus of this chapter, the need to assess these parameters should be considered in concert with the development of the wind resource assessment campaign, as they could affect the siting, design, and operation of offshore monitoring stations. For example, platforms might have to be larger to accommodate ocean- and wildlife-monitoring equipment, they might need a more robust power supply as well as additional data logging and communications equipment, and they might require a different schedule of operations and maintenance visits. All these factors can increase capital and operations and maintenance costs.

Finally, the decision of where to place an onshore monitoring station should take into account the potential value of collecting data throughout the project's life cycle. Placing the primary monitoring station within the planned array, as is usually done for onshore projects to improve the accuracy of wind flow modeling, implies that any wind resource data collected after the turbines become operational will be affected by turbine wakes, making it difficult to obtain an accurate measure of the free-stream speed. Siting the station upwind of the planned array, on the other hand, could yield useful data for the ongoing analysis of the plant's performance, without substantially reducing the accuracy of the preconstruction assessment because the wind resource varies relatively little offshore. Given the costs, challenges, and time required to carry out an offshore site assessment campaign, it is especially important to consider the full scope of data needs for the project, both pre- and postconstruction, from the outset. Coordinating and synthesizing the wind project's design, permitting, engineering, and operational data needs will result in a comprehensive and high value monitoring campaign.

14.1 NATURE OF THE OFFSHORE WIND ENVIRONMENT

Offshore wind environments differ from onshore environments in a number of ways. One difference is that the surface roughness (which determines the drag exerted by the surface on the lowest layer of the atmosphere) of open water is much smaller than that of most land surfaces. A typical roughness length assumed in numerical modeling is 0.001 m, although the value varies with wave height and therefore with wind speed. In contrast, most land surfaces have a roughness ranging from 0.03 m to over 1.0 m (Table 13-1). The low roughness of water means that the wind shear offshore tends to be lower than that observed on land. Average wind shear exponents over water typically range from 0.07 to 0.15, compared to 0.10 to 0.60 over land (Table 10-3). Turbulence is generally lower as well.

Another difference is that the daily cycle of surface temperature variation is usually attenuated offshore because water has a much greater heat capacity than soil and maintains a more constant temperature throughout the day. This characteristic produces, in turn, smaller variations in atmospheric stability and wind shear. Whereas on land, the mean wind shear can vary greatly between night and day (Fig. 11-3), such patterns are not usually as evident offshore. In general, the average shear exponent is lower in tropical waters (0.07–0.10) than that in temperate and cold waters (0.10–0.15). This is because in the tropics, the water is warm and the atmosphere close to neutrally stable year-round. In colder climates, seasonal variations in the relative temperature of air and water modify the stability and shear, producing higher average shears on the whole.

Because of the lack of terrain, winds and other meteorological conditions tend to be more spatially uniform offshore, especially farther than around 5 km from the land. This is fortunate for wind project development, as it means that fewer measurement stations are generally required to characterize the resource accurately within a project area. Even so, surprisingly complex wind phenomena can occur. Here are some examples.

- *Mountain and Island Blocking.* Coastal mountains and islands can act as a barrier creating a zone of low wind speeds both upwind and downwind. This effect can extend many kilometers offshore depending on the atmospheric conditions and the size of the barrier. Figure 14-1 shows an example of blocking by mountains on the island of Maui, Hawaii, USA.
- *Gap Flows.* Gaps between and around coastal mountains and islands can concentrate the wind and generate high wind speeds offshore. Figure 14-1 shows such channeling between the islands of Maui, Lanai, and Molokai. Many other examples exist, some with familiar names such as the Mistral off the coast of southern France and the Levant through the Strait of Gibraltar.
- *Coastal Barrier Jets.* When synoptic conditions favor a flow more or less parallel to the coastline, the terrain can act to concentrate the flow and create a low level jet with high wind speeds. Figure 14-2 shows a numerical simulation and a synthetic aperture radar (SAR) image of such a jet off the north shore of the St. Lawrence River in Canada.

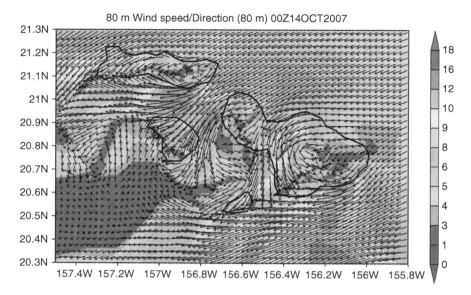

Figure 14-1. A numerical simulation of wind speeds and directions at 80 m height around the islands of Maui, Lanai, and Molokai, in the Hawaiian Islands, showing the effects of both mountain blocking (blue and purple areas) and channeling (red areas). Note how the wind direction, indicated by the arrows, is deflected by high pressure on the upwind side and by low pressure on the lee side of the islands. *Source:* AWS Truepower.

- *Roughness Transitions.* When the wind comes off the land, the abrupt decrease in roughness generates a zone of gradually increasing wind speed near the surface, called an *internal boundary layer* (IBL), whose depth grows with distance offshore (Fig. 14-3). Above this IBL, the original wind profile is largely unaffected. Depending on the wind direction, distance from shore, and rate of growth of the IBL, the transition may occur either above or below the hub height of the turbines in an offshore wind project.

- *Stability Transitions.* In addition to a decrease in roughness, wind coming off the land can sometimes encounter a large difference in surface temperature, which produces changes in thermal stability. For example, if warm air moves over cooler water, as often occurs on summer days in the middle and high latitudes, the boundary layer becomes thermally stable. This can cause the atmosphere to decouple from the surface layer, allowing strong winds to build at heights near the hub heights of wind turbines.

- *Mesoscale Circulations.* Surface temperature and moisture gradients can create mesoscale wind circulations, which can affect the offshore resource. A classic example of a temperature-driven circulation is a sea or lake breeze (Fig. 14-4). On a typical summer day, as the sun heats the land surface, the air above it tends to warm and rise, causing relatively cool, moist air to be pulled in from over the water. (The opposite circulation, a land breeze, can occur at night as

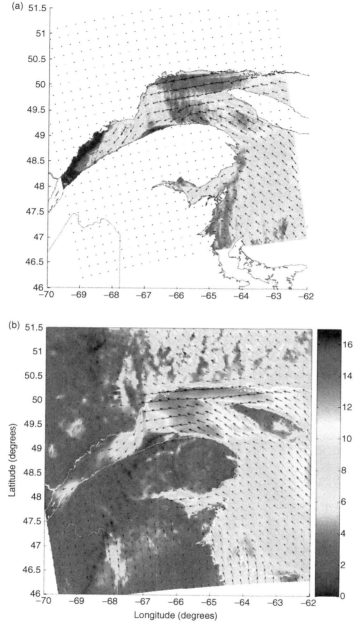

Figure 14-2. A synthetic aperture radar image (a) and a numerical simulation (b) showing wind speeds and directions at 10 m height in the Saint Lawrence River for the same date and time. The red area off the northern shore in both images is a coastal barrier jet. *Source:* Beaucage et al., 2007 (1).

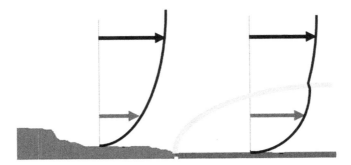

<u>Figure 14-3.</u> Schematic representation of the evolution of a wind speed profile as the wind moves off the land over water (from left to right). The initial profile reflects the land surface roughness. As the air moves over the much smoother water, an IBL (yellow line) develops and grows with distance from shore. Within the IBL, the wind speed increases, and the wind profile assumes an offshore shear pattern. *Source:* AWS Truepower.

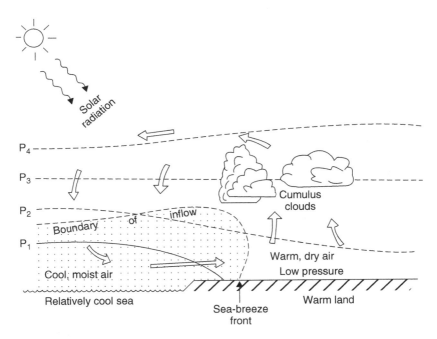

<u>Figure 14-4.</u> A schematic representation of a sea breeze circulation. As the sun heats the land, the air above it warms and rises (right). This produces a drop in pressure, which draws cool air from over the water. *Source:* Royal Meteorological Society.

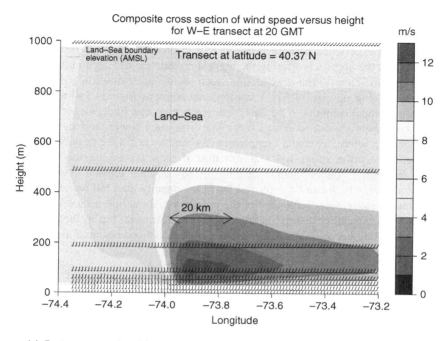

Figure 14-5. A cross-sectional image of mean wind speeds in meters per second as a function of height (vertical axis) and longitude from west to east (horizontal axis) off the New Jersey (USA) coast from a numerical simulation of a composite of 40 summer days at 4 PM local time. The simulation shows that when such sea-breeze-induced jets form, wind speeds above 10 m/s can extend from about 50–300 m in height from near the coast up to several tens of kilometers offshore. *Source:* Freedman et al., 2010 (4).

the land cools, but it is usually less pronounced.) In the absence of a strong background wind flow, a sea-breeze front can progress as much as 50 km inland from the coastline (2). When the synoptic flow reinforces the sea breeze circulation, the result can be a high speed, low level jet. Research suggests, for example, that such a jet appears periodically off the coasts of New York and New Jersey, in the United States, during the warm season. The large zone of intense winds that is formed could benefit offshore wind energy production in this region (Fig. 14-5) (3).

14.2 WIND RESOURCE MONITORING SYSTEMS

The offshore wind industry has yet to establish a standard approach to offshore wind resource assessment. Even in Europe, where wind plants have operated offshore for over 20 years, the projects have been financed based on a variety of measurement methods, including hub-height masts equipped with anemometers, stand-alone lidars, and, in some cases, no on-site measurements at all.

This mixed experience notwithstanding, a fixed, instrumented tall tower must still be regarded as the approach offering the highest confidence and accuracy. For reasons that will become clear in the next section, such towers are extraordinarily expensive, with costs ranging upward of $2–3 million, and some as high as $10 million. They can also take many months to be designed, permitted, and built. Consequently, few project developers choose to install more than one such tower, and there is always keen interest in alternatives and complementary approaches. One possible alternative is to install the tower on an existing structure, if there is one not too far from the project area. Other options include floating or fixed remote sensing systems (mainly lidar and sodar), which can be deployed far more cheaply than tall towers, as well as satellite-based radar measurements. For the time being, these options are best regarded as complementing rather than replacing purpose-built towers, but in the future, lidar, in particular, may become a primary instrument for offshore resource assessment.

14.2.1   Purpose-Built Meteorological Towers

To date, the majority of European offshore meteorological towers for resource assessment have been purpose-built, self-supporting lattice structures, such as that shown in Figure 14-6. To understand why such a tower should be so expensive, consider that the tower height is the sum of the heights above and below water. Thus, a tower deployed in water 30 m deep and extending 80 m above the water is actually 110 m tall from the seabed to the top. Moreover, the part in contact with the water must be able to withstand powerful ocean currents and waves. Water is about 800 times denser than air, and so a typical tidal current of, say, 2 m/s exerts a force equivalent to a wind speed of 57 m/s—a class 3 hurricane. During storms, waves can exert still larger forces, and in cold climates, the impact of ice must also be accounted for in the structural design. Add to that the potential wind and ice loads on the parts of the structure that are above the water, and the result is that offshore masts must be massive and their foundations attached securely to the sea floor.

Offshore wind monitoring towers are typically equipped with instruments very similar to those deployed on land, including anemometers, direction vanes, and air temperature and pressure sensors, although marine-grade models may be called for. In addition, the towers and their platforms may have motion sensors to detect deflections caused by wind and waves, which can affect speed measurements, and a variety of instruments to measure ocean currents, wave heights, water temperature, and other parameters-even bird activity. These other elements can transform an offshore mast into a complete ocean-atmosphere-wildlife monitoring system, providing data critical to the design and permitting of a wind project.

The anemometer configuration on an offshore mast generally differs somewhat from that recommended for land-based towers. Typically three or more anemometers (at least one per tower face) are installed at each monitoring height, rather than only two. This added redundancy aims to achieve a high overall data recovery with fewer visits for maintenance and repair. Furthermore, having three anemometers (installed

Figure 14-6. (a) Offshore meteorological tower for the NoordzeeWind project.

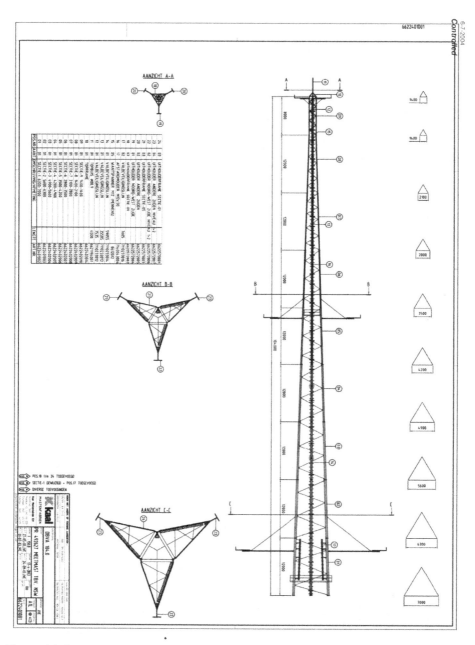

Figure 14-6. (b) Diagram of the NoordzeeWind met tower. The section views of the boom assemblies along the left side illustrates the design of the reinforced boom structures. *Source:* Eecen PJ, Branlard E. The OWEZ Meteorological Mast. Analysis of mast-top displacements. NoordzeeWind report. Report OWEZ_R_121_mast_top_movement.pdf.

120° apart) at each height allows for the wind to be measured by at least two sensors from every direction, thereby reducing the influence of the large tower on the wind measurements.

One of the main challenges associated with self-supporting offshore masts is their large face width. In order to minimize the effect of flow distortion on the measurements, IEC guidelines recommend a minimum separation distance on lattice towers of three to four face widths, depending on structure porosity. Achieving this standard may be impossible at every monitoring level. Nonetheless, adequate instrument separation requires booms that are significantly longer than those typically installed on land-based met towers. These must be reinforced, as shown in Figure 14-6, to maintain structural stability. The booms in this illustration are designed to rotate vertically toward the tower so that the instrumentation at the end of the boom can be accessed for maintenance. The booms are held in place by tensioned wires connected to the tower.

The design and deployment of an offshore meteorological tower usually takes much longer than that of a land-based tower. In general, several preconstruction studies are required to obtain the necessary permits. A variety of meteorological, oceanographic, and geophysical studies are needed to characterize the environmental conditions the tower will face as well as the geological conditions below the sea or lake floor.

As shown in Figure 14-7, the installation of the FINO 1 research platform, an offshore meteorological tower in the North Sea, required a variety of vessels and equipment. For this particular installation, the jacket foundation structure was first lowered onto the seabed and piles were then driven through the foundation's pile sleeves and into the seabed using a hydraulic hammer. The piles and foundation were bonded using grout. Next, the platform, on which the met tower had been partially constructed, was installed by a crane barge and grouted into place. Finally, the remaining sections of the met tower were installed by the jack-up barge's crane.

The FINO 1 tower represents just one way that offshore towers can be constructed. Many other approaches can be taken depending on the characteristics of the platform and tower. For example, lattice towers can be erected using a gin pole, which extends above the top of the partially built tower and is equipped with a pulley, allowing the next section of the tower to be lifted into place.

Offshore towers will most likely be deflected to some extent by wind and waves during operation. Although previous studies have found that such movements have only a small impact on wind speed measurements (5), the impacts for any particular mast depend on the station design and instrument choice. Thus, it is prudent to measure the deflections directly to be able to quantify the impact on speed. The installation of multi-axis accelerometers at each monitoring height should be considered to accomplish this.

14.2.2 Surface-Based Remote Sensing Systems

As on land, surface-based remote sensing technologies, mainly lidar and sodar, are a useful complement to traditional tower-based measurement campaigns offshore. Colocated with a mast, they can extend the wind resource measurements from the mast height up to the top of the turbine rotor plane. As the use and acceptance of remote

Figure 14-7. Installation of the FINO 1 offshore research platform using a jack-up barge (left) and a crane barge (right). *Source:* FINO 1 research platform, Copyright Bilfinger Berger Civil, Germany.

sensing advance, offshore developers are increasingly turning to these devices (primarily lidar) as stand-alone wind monitoring tools. This trend is reinforced by the growing use of remote sensing for such demanding applications as power curve verification. Since lidar and sodar systems are much smaller and lighter than tall towers, they can usually be deployed at a far lower cost. For this reason, it is not uncommon for them to be used to provide a first indication of the wind resource in a project area, before the developer decides to invest in a purpose-built tower.

This section provides an overview of the various applications of surface-based remote sensing, including vertical and side-scanning lidar and sodar. Satellite-based radar, another remote sensing technology, is discussed later in this chapter.

Vertical Profiling Lidar. The use of vertical profiling lidar units offshore has grown in recent years. If anything, lidar is more widely accepted offshore than on land because the lack of terrain and low surface roughness allow for even closer agreement with cup anemometers, the standard for defining wind turbine power curves.

Lidars can be deployed on many types of fixed platform, including jack-up barges. They must be carefully placed and configured to ensure good accuracy and data quality, however. The specific requirements depend in part on the type of system. In general, a lidar should be placed where other platform elements, such as towers, maintenance buildings, and moving equipment such as anemometers and wind vanes, will not create interference. It also needs to be accessible for periodic maintenance. Additionally, some lidar units use onboard meteorological sensors as an input to their data processing, and it is important that they be well exposed to the environment.

Although it is cheaper to deploy a lidar system on a fixed platform than to install a tall tower, it can nonetheless be an expensive and time-consuming process. A promising alternative approach is to place the lidar on a floating platform or buoy

(Figure 14-8). In theory, this could be far less expensive than a fixed platform, and it could also be deployed in much deeper water with fewer permitting hurdles. Like mobile lidars on land, a floating lidar could be moved around for site prospecting and spot-checks of the resource, and it could be deployed for performance measurements at operating wind projects. When coupled with likely advancements in lidar technology and industry acceptance, the floating lidar concept could greatly benefit offshore resource assessment.

Several companies are exploring floating lidars, and a few systems are commercially available. However, given the challenges associated with this technology (including motion compensation or correction, power supply integration, and overall system reliability), it is likely that these systems will require considerable validation and testing before they can be fully accepted as equivalent to tower-based measurements.

Side-Scanning Lidar. A side-scanning lidar operates on the same physical principles as a vertical profiling lidar, but as the name suggests, the laser beam can be pointed sideways as well, enabling it to scan over a hemisphere. The potential advantage of this technology for offshore resource assessment is that a side-scanning lidar with enough range could be deployed on land or on a convenient offshore structure

Figure 14-8. The WindSentinel floating lidar system consisting of a Vindicator lidar from Catch the Wind, mounted on a NOMAD buoy developed by AXYS Technologies. *Source:* AXYS Technologies.

outside the project area. From this vantage point, the unit could produce a three-dimensional grid of wind velocities up to and above the turbine hub height throughout the area. The current maximum range of commercially available side-scanning lidar units, which varies from 4 to 15 km, is promising. The use of side-scanning lidars for wind resource assessment remains a relatively new application of this technology, however. Consequently, additional validation will likely be necessary before it is fully accepted for resource assessment.

Sodar. Sodar systems have been deployed offshore on fixed platforms both for wind resource assessment and wind energy research (Figure 14-9). Sodar offers some attractive attributes for offshore operation, such as moderate power requirements, good reliability, and relatively low cost. However, sodar can be susceptible to acoustic interference. Data quality can be affected by the sound of wind whistling through structural components, audible navigation aids, and noise generated by waves and wildlife. Consequently, commercial floating versions of sodar have not been developed to the extent as lidar. While sodar may be a viable choice for some offshore measurement campaigns, lidar's dominance is expected to continue to grow.

14.2.3   Using Existing Offshore Structures

In some instances, it may be possible to take advantage of existing structures to support offshore wind monitoring towers and remote sensing systems. Provided wind conditions at the site are reasonably representative of conditions in the proposed project area, and provided the structure can support the desired monitoring system without interfering in its operation, the use of existing offshore facilities can add value to a resource assessment campaign at a much lower cost than a purpose-built platform. Lighthouses on exposed points, C-MAN (Coastal-Marine Automated Network) stations in the United States and similar stations elsewhere, oil-drilling platforms, and even small, low lying islands can be considered for this application.

<u>Figure 14-9.</u> (a) Scintec sodar unit mounted on the Ambrose Light Coastal-Marine Automated Network (C-MAN) station off the coast of New York, USA. *Source:* AWS Truepower. (b) A Second Wind sodar unit being lifted onto a U.S. Coast Guard platform off the coast of South Carolina, USA. *Source:* Second Wind.

<u>Figure 14-10.</u> (a) Monitoring tower installed on the top of the Cleveland Crib facility. *Source:* EcoWatch, Ohio. (b) Ambrose Light C-MAN station outfitted with a suite of wind resource assessment equipment. *Source:* AWS Truepower.

One example of where this approach has been undertaken is the Cleveland Crib, which houses the main municipal water supply intake for the city of Cleveland, Ohio, USA, on the shore of Lake Erie (Fig. 14-10a). A 30-m monopole tower, equipped with three levels of instrumentation, was installed on the top of the structure in 2005 to support the development of an offshore wind farm. More recently, the facility has also hosted a lidar unit.

Another example is the Ambrose Light C-MAN station off the coast of Long Island, New York, USA (Fig. 14-10b). Owned and maintained by the US National Data Buoy Center, this C-MAN station was outfitted with anemometers and a sodar system to help characterize the offshore wind environment in the Atlantic Ocean off the coasts of the states of New York and New Jersey.

Often, an important challenge of using an existing structure is that it may influence the wind measurements. Large structures such as lighthouses, oil-drilling platforms, and the like cause substantial flow distortions all around them, either reducing or increasing the observed speeds depending on their shape, the wind direction, and the placement of the wind measurement system. These distortions can sometimes be estimated and partially removed through CFD modeling, but nevertheless, the accuracy of the corrected measurements usually suffers.

14.2.4   Weather Buoys

Weather buoys can be useful for offshore resource assessment in several ways. First, they can be deployed early in the development process, with fewer permitting hurdles than fixed platforms, and thus can support preliminary site assessments and help establish a longer baseline of surface meteorological and ocean measurements.[1] Later, if a fixed platform is deployed, the measurements from both buoy and platform can be integrated into a more complete assessment of the ocean-wind environment.

---

[1] It should be noted that in environments subject to severe icing, weather buoys might have to be retrieved during the winter. The resulting gaps in the data stream could offset some of their advantages.

Second, weather buoys can serve as the primary on-site data-gathering station for ocean parameters, including waves, currents, water levels, temperatures, salinity, and others, relieving the need to measure these parameters on a fixed platform. Last, following the deployment of fixed, primary monitoring stations, one or more weather buoys can be deployed (or redeployed) throughout the project area to supplement the wind resource assessment campaign.

The buoy's basic atmospheric monitoring package should include wind speed and direction (using redundant sensors at least 3 m above the water surface), barometric pressure, and air and water surface temperature. The buoy should include an onboard data acquisition system, power supply, and a communication package capable of transmitting data at least as often as every 2 weeks. Including a GPS position reading in the transmitted data will help the owner track the buoy in the event it comes loose from its mooring and drifts.

14.2.5   Data Logging and Communications Systems

Data logging and transfer are performed offshore in much the way they are on land: data are collected and stored temporarily in a data logger, then retrieved remotely, or if that is not possible, during site visits. Because of the cost and time required to visit an offshore platform, it is recommended that additional redundancy be provided to make the data collection and remote transfer as reliable as possible. A second, parallel data logging system should be considered as a backup. It is also recommended that a computer, or other redundant data storage device, be placed on-site to provide backup storage in case of an extended communications failure. If a computer is used, it should connect the different monitoring systems in one network. If equipped with redundant, high capacity data storage (e.g., a disk array), the computer can serve as a reliable repository for data from all the networked devices. In addition, if properly equipped, the computer could provide a redundant means for remote users to interface with the devices.

Highly reliable communications are needed to transmit and receive the data and monitor the status of equipment deployed offshore. Accordingly, a redundant remote communication and control system is recommended for all offshore monitoring equipment. The communications system must also have sufficient bandwidth, or speed, to support the required volume of data to be transferred. Given that many offshore project sites are located outside the coverage areas of cellular networks, cellular communications may not be feasible. A satellite connection is a possibility, but it may not be adequate for transferring large data files because of bandwidth constraints, data-transfer caps, and cost.

Often, a point-to-point radio connection is the best option for transferring large data files and providing rapid (low latency) remote access to offshore systems. This technique, which is rarely employed for land-based resource assessment, uses a pair of directional antennas pointed directly at one another to connect a base station computer on land to a remote computer on the offshore platform (Fig. 14-11). Operating on licensed and unlicensed RF bands, the antennas form a high bandwidth wireless network connection between the two computers. Files can be transferred from the

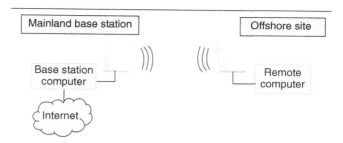

<u>Figure 14-11.</u> Diagram of a possible point-to-point radio connection between an offshore monitoring station and a computer on land. *Source*: AWS Truepower.

remote computer quickly, and unlike commercial cellular- or satellite-based connections, there are no data-transfer limitations other than those imposed by the equipment. The method requires that the base station be within the range of the offshore antenna and have a clear line of sight to it.

14.2.6 Power Supply

Most offshore platforms do not have access to grid power, and therefore, an independent power supply is usually required. This can place constraints on the amount and types of monitoring equipment that can be supported, especially considering that the platform area, loading considerations, and other equipment can limit the space available for the power supply. Most offshore power supplies consist of a generation system (solar panels, small wind turbines, fuel cell, a generator, or some combination of these) and a bank of deep-cycle storage batteries. The design capacity and configuration of the power supply are influenced by the equipment attached to it, including sensors, data loggers, and communications systems, as well as by the desired transmission frequency, uptime, and length of time the system must operate unattended. In some cases, the power supply can also affect permitting, especially when diesel, propane, or other fuels need to be transferred to the platform and stored on it.

The standard instruments deployed on towers and their loggers consume relatively little power. A bigger challenge is posed by remote sensing systems, especially lidar. Although manufacturers continue to make progress on reducing the power consumption of their units, lidars can still represent a significant power requirement for offshore platforms. The power needs for lidars vary by device and operating conditions. Typically, during periods of extremely cold or warm weather, a lidar's onboard temperature control system will draw more power. Some lidar units have thermally insulated "jackets" that can be installed in the winter to help regulate the unit's temperature.

Aside from the wind resource assessment systems, other equipment on the platform, such as warning lights, navigation aids, biological monitoring devices (e.g., radars), and ocean environment sensors, must be considered in determining the overall power requirements.

14.3 OPERATIONS AND MAINTENANCE OF OFFSHORE SYSTEMS

No matter what monitoring systems and platforms are used, confidence in the data collected in a resource assessment campaign depends on having an effective operations and maintenance program. The fundamentals of such programs are the same for offshore and onshore applications. However, the offshore environment makes operations and maintenance a more challenging and costly task. The following section focuses on how the operations and maintenance of offshore stations differ from typical practices for land-based stations.

14.3.1 Site Visits

Visits to offshore monitoring stations are usually conducted in small- to medium-sized workboats, depending on the scope of the activity to be performed. During the winter, in cold climates, it may be necessary to use an ice-breaking vessel. When visiting large stations, and especially far offshore, it may be wise to use a helicopter (assuming the station is equipped with a suitable helicopter landing pad). Helicopters reduce travel time and can therefore increase the time on site.

Whatever method is used, the weather is always a concern, and it is essential that weather and sea-state forecasts be consulted when planning offshore site visits. In general, the maximum sea state for accessing offshore platforms from a workboat is about 1.5 m; the wave tolerance can be even less when personnel or equipment need to be transferred between the vessel and a platform or other offshore structure. Helicopters fly above the waves but are susceptible to high winds. For these reasons, it can be difficult to find a window of opportunity for an extended visit during those times of year that have the strongest winds and greatest storm activity (winter, in most offshore wind development areas). If possible, major maintenance activities should be scheduled during the months when the wind and sea state are normally calmer.

14.3.2 Operations and Maintenance Procedures

Many of the maintenance procedures for offshore monitoring stations are the same as those for onshore stations, although they can be more complicated and expensive to carry out. For one thing, most offshore towers cannot be tilted down like the guyed, tilt-up towers commonly used on land. This means that certified tower climbers must perform all the up-tower work. Their ability to do this safely and effectively depends on the site conditions.

The amount of routine on-site maintenance that is needed depends on the monitoring equipment. Self-supporting masts with standard meteorological sensors and data loggers generally require the least maintenance. As is the case on land, it is recommended that wind vanes and cup anemometers be replaced on a regular basis (no less often than every 24 months), as part of a preventive maintenance program. Both lidars and sodars typically require more frequent maintenance and repair visits than do towers, although it can be expected that as manufacturers deploy systems specifically adapted for harsh and remote offshore environments, the time between visits will increase. Scheduled lidar and sodar maintenance activities can include replenishing window-washing fluid, changing wiper blades, removing bird droppings, and replacing desiccant to control humidity.

Offshore power supplies also require periodic maintenance. On unmanned platforms, birds have a tendency to congregate on all available perches. For this reason, where solar panels are used, their output should be monitored and compared to readings from a pyranometer to determine if the panels are becoming excessively soiled. It is recommended that bird deterrents be considered for all critical equipment. If they are not implemented or if they prove ineffective, it may be necessary to manually wash off solar panels and other equipment from time to time. When a generator is part of the power supply system, serving as either a primary or an emergency backup power source, it will require not only refueling but also periodic maintenance such as oil changes and perhaps engine overhauls, depending on how often it is used. Precautions should be taken if the storage batteries in the power supply are housed in a confined space, as they may pose an air quality hazard for which additional safety training and equipment (such as an air quality monitor) may be needed.

During operation, it is important to monitor the health of the sensors by analyzing their data on a regular basis. Furthermore, it is recommended that status indicators from ancillary equipment on the tower, such as warning lights and power supply, be included in the data transmitted from the station. This information can be monitored to verify that the systems are operating normally and to support maintenance planning.

Because of the cost and effort needed for a maintenance visit to an offshore monitoring station, it is recommended that ample spare parts for critical equipment be brought along during maintenance visits. This is particularly true for equipment that typically has an extended replacement lead time, such as calibrated anemometers.

14.4 SATELLITE-BASED MICROWAVE SENSORS

Given the costs and inconveniences of surface-based wind monitoring systems, it is natural that there should be a strong interest in other potential sources of offshore wind resource measurements. The leading candidates are satellite-based microwave sensors, and in particular, microwave radiometers, scatterometers (SCATs), and SARs. All these instruments operate on the principle that winds create waves, which affect the roughness of the sea surface. The degree of roughness, and specifically the size distribution of the small wavelets[2] that form almost instantly in response to the wind, is in turn related to the amount of microwaves emitted or reflected from the surface at different frequencies. Using mainly speed measurements from weather buoys as the standard

---

[2]The signal detected by microwave sensors is most sensitive to waves that are about the same size as the microwave wavelength. This ranges from about a millimeter to several centimeters for most satellite-based microwave instruments.

of comparison, researchers have developed semiempirical relationships between measured wind speeds and microwave observations of various types and frequency bands. As a result of this research, over the last two decades, a large amount of offshore surface wind speed data from a variety of instruments have become available (6).

A key advantage of using microwaves instead of other parts of the electromagnetic spectrum is that they are indifferent to day and night and, depending on the frequency, can pass through most clouds. Also, unlike cloud-tracking satellites, satellite-based microwave sensors measure winds near the surface. However, the areal coverage, spatial resolution, and revisit frequency[3] of the satellite-based sensors vary greatly. Radiometers and SCATs tend to cover large areas with greater frequency than SAR, but their spatial resolution is coarse. SAR achieves a much higher spatial resolution, but with fewer SAR images available, its temporal coverage is more limited.

Because they do not measure speed directly but rather infer it from properties of the sea surface, it is unlikely that satellite-based microwave observations will achieve the accuracy needed to replace on-site towers, lidar, or sodar for resource assessment. One important limitation is that the microwave signals are calibrated to buoy measurements taken just a few meters off the surface, which are adjusted to a nominal 10 m height. The resulting speed estimates must then be projected to the hub height of wind turbines. Each stage of this process involves approximations and assumptions that can affect accuracy. Nevertheless, in somewhat the manner of numerical atmospheric models, satellite-based microwave data can be an excellent tool for site prospecting and, in conjunction with high quality surface-based wind measurements, for characterizing the variation of the wind resource within project areas.

14.4.1 Types of Sensors

Table 14-1 lists many of the wide variety of satellite-based microwave sensors that have been deployed. The list changes often as new satellites with improved sensors are launched almost every year. The three main types of instrument, namely radiometers, SCATs, and SARs, are described in the following sections.

Passive Microwave Radiometers. Passive microwave radiometers, such as SSM/I, TMI, AMRS-E, and WindSat, record emissions from the earth's surface in a number of frequency channels. (They are called *passive* because, unlike radar, they do not bounce signals off the earth's surface but only passively observe its emissions.) From these measurements, daily global wind maps are produced at a horizontal resolution of approximately 25 km. These maps, in turn, can be compiled into maps of monthly, seasonal, and annual mean wind speeds. Because the accuracy of the speed estimates is affected by the presence of land within the sensor's field of view, valid observations are unfortunately limited within about 25 km of coastlines, where most offshore wind projects are located.

---

[3]The revisit frequency is the number of times per day a satellite passes within range of the same location.

Table 14-1. Satellite-based microwave sensors providing sea-surface wind speed data

Satellites	Agency	Year of launch	Instrument	Radar band	Resolution
SSM/I	NASA	1987	Radiometer[a]	K, Q, W	25 km
RADARSAR-1	CSA	1995	SAR[b]	C	~25 m
ERS-2/SAR	ESA	1995	SAR[b]	C	~25 m
TMI	NASA–JAXA	1997	Radiometer[a]	X, K, Q, W	25 km
Quick SCAT/Sea Winds	NASA	1999	SCAT	Ku	25 km
ENVISAT/ASAR	ESA	2002	SAR[b]	C	~30 m
AMSR-E	JAXA	2002	Radiometer[a]	C, X, K, Q, W	25 km
WindSat	NASA	2003	Radiometer[a]	K, Q	25 km
ASCAT/METOP-1	ESA	2006	SCAT	C	25 km
ALOS	JAXA	2006	SAR[b]	L	~25 m
RADARSAT-2	CSA	2007	SAR[b]	C	~25 m

Abbreviations: CSA, Canadian Space Agency; ESA, European Space Agency; JAXA, Japan Aerospace Exploration Agency.
[a] Passive microwave radiometers usually record the earth surface emissivity at several different frequencies.
[b] Space-borne SARs have different beam modes, each with different pixel sizes.
Source: Beaucage et al., 2008 (7).

Scatterometers. A microwave SCAT is a type of radar. It consists of a transmitter, which emits microwave pulses toward the earth's surface, and a receiver, which receives their echoes. (In most SCATs, the transmitter and receiver employ the same antenna.) An empirical relationship relates the backscattered signals to the wind speed at 10 m height. Rainfall can contaminate the signals from SCATs, especially those that operate at smaller wavelengths (higher frequencies), but a rain flag provided with the data allows the analyst to remove such contamination. The leading SCATs cover more than 90% of the ocean surface on a daily basis at a resolution of about 25 km. Like passive microwave radiometers, their coverage and quality are limited near shorelines. Unlike radiometers, however, SCATs can also measure the wind direction, an important advantage for resource assessment.

Synthetic Aperture Radars (SARs). While SCATs were designed expressly to map global ocean winds, SAR systems were built for a variety of research purposes, such as ice characterization and mapping, oil spill detection, ship detection, and wind and wave measurements (8). SARs analyze the same characteristics of the backscatter signal as do SCATs, but they can achieve much finer spatial resolution (as high as 10 m) through signal processing tricks that allow the moving transmitter/receiver to mimic a much larger antenna. Like SCATs, SARs use an empirical relationship to derive the surface wind speeds from the backscattered signals. In contrast to SCATs, however, SAR systems cannot measure wind direction directly, although it can sometimes be inferred from the presence of atmospheric roll vortices or wind streaks. More commonly, the analyst must obtain directional data from another source (such as surface-based measurements, another satellite-based sensor, or a numerical weather model).

Like other satellite-based microwave sensors, SAR systems provide a snapshot (or image) of the near-surface winds at a particular time. Since the image retrievals must be requested from the multipurpose sensor, the number of archived SAR images varies from one region to another. Furthermore, because they cover a smaller area (a few hundred kilometers wide at most), SAR satellite passes over the same location are less frequent than those of satellite-based SCATs and radiometers. One consequence is that SAR systems, unlike SCATs and radiometers, do not provide a gridded data set, and some expertise in radar image processing is needed to derive the wind speeds from the raw images.

14.4.2 Wind Resource Assessment Using Satellite-Based Microwave Sensors

Wind resource maps from radiometers and SCATs can provide a useful initial indication of the wind resource in a particular region. Among their advantages are that they are available for free and they offer global coverage with daily or more frequent satellite passes. An extensive archived data set from the SSM/I radiometer is available since 1987. However, the wind speeds derived from SCATs such as QuikSCAT are considered to be more accurate than those derived from radiometers. While the period of record of the QuikSCAT data set, 1999–2009,[4] is not as long as that of SSM/I, it is sufficient to obtain fairly reliable estimates of long-term mean wind speeds. Several studies have shown that the instantaneous wind vector accuracy (root-mean-square error) for QuikSCAT data is around 1.0–2.0 m/s, which may be somewhat better than that of numerical weather prediction models but is worse than that of anemometers, lidar, or sodar. Figure 14-12 shows two examples of the kinds of global ocean surface wind resource maps that can be obtained from the QuikSCAT data.

Satellite-based SAR imagery, on the other hand, offers considerable promise for project-scale resource assessment, as it can achieve a much better resolution than radiometers and SCATs and can also measure winds near coastlines (as demonstrated in the SAR image in Figure 14-2a, which shows complex near-shore wind patterns in the St. Lawrence River). Although the spatial resolution of SAR is in principle very fine indeed, in practice, problems such as speckle limit the grid spacing for most applications to a few hundred meters. This is nonetheless comparable to the spacing between wind turbines in a wind project. Thus, SAR can provide information about the wind resource distribution of practical use for designing wind projects.

Several studies have shown that wind speeds derived from SAR are about as accurate as SCAT-derived speeds when compared to buoy measurements. Other than the inherent limitations of the method, the utility of SAR-derived wind speed data depends mostly on the number of SAR images available for a particular site of

---

[4]NASA PO.DAAC web site: http://podaac.jpl.nasa.gov/. Ifremer web site: http://www.ifremer.fr/cersat/en/data/overview/gridded/mwfqscat.htm. QuikSCAT stopped operating in November 2009.

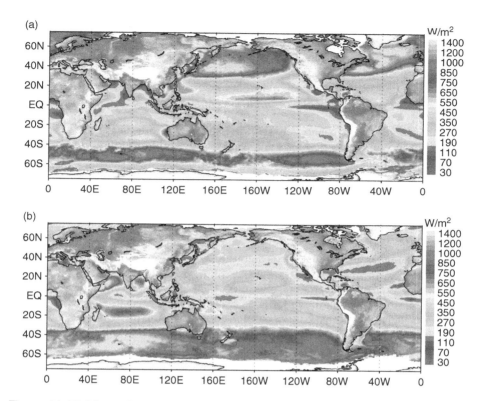

Figure 14-12. Maps of mean wind power density derived from QuikSCAT for (a) boreal winter (December, January, and February) and (b) boreal summer (June, July, and August), for the 8-year period between 2000 and 2007. The gray scale corresponds to topography. *Source:* Liu et al., 2008 (9).

interest.[5] Unfortunately, SAR images are expensive. The commercial price for a single image ranges from a few hundred to a few thousand US dollars, depending on the satellite and distributor, whereas several dozen to hundreds of images are required for a statistically reliable estimate of the annual mean wind speed. (Images stored in archives are significantly cheaper, and some may be available for free through government institutes or research projects sponsored by the European Space Agency or Canadian Space Agency, for instance.)

For the foreseeable future, SAR is likely to be used as a complement to, rather than replacement for, on-site measurements and numerical wind flow modeling. Its importance is likely to grow, however, if SAR image prices come down and methods of processing large numbers of images improve. One potentially useful application of

---

[5]For a well-distributed sample of SAR images, the total uncertainty on the mean wind speed decreases approximately as one over the square root of the number of images.

SAR is mapping wakes from operating offshore wind projects. The extent of wake influences offshore is an important issue, as very large wind projects are planned in areas such as the North Sea, and realistic wake simulations remain a difficult challenge for numerical models.

In interpreting wind speeds derived from all satellite-based microwave sensors, it is important to keep the following in mind:

- Microwave sensors cannot retrieve wind speeds when the sea surface is covered by ice, which means that for some parts of the world, the winter season can be underrepresented in the image or data sample.
- The wind speeds represent spatial averages over the sensor's effective resolution for a particular moment in time, whereas anemometers and surface-based remote sensing systems typically measure time-averaged winds for a specific point. This difference means that satellite-based measurements rarely match surface-based measurements exactly.
- Most satellites are in polar orbits that take them over the same location at specific times each day. For instance, the QuikSCAT satellite passed twice daily over Eastern Canada at approximately 10:00 UTC and 22:00 UTC. It is, therefore, not possible to obtain the full diurnal cycle of wind speeds from those satellites. Fortunately, the diurnal pattern of wind speed variation offshore is usually much less than it is on land, and therefore, the errors introduced by this limited sampling are generally minor.

REFERENCES

1. Beaucage P, Glazer A, Choisnard J, Yu W, Bernier M, Benoit R, Lafrance G. Wind assessment in a coastal environment using the synthetic aperture radar satellite imagery and a numerical weather prediction model. Can J Remote Sens 2007;33:368–377.
2. Stull RB. Introduction to boundary layer meteorology. USA: Kluwer Academic Publishers; 1988. p. 666.
3. Colle BA, Novak DR. The New York Bight jet: climatology and dynamical evolution. Mon Weather Rev 2010;138:2385–2404.
4. Freedman JF, Bailey B, Young S, Zack J, Manobianco J, Alonge CJ, Brower M. Offshore wind power production and the sea breeze circulation (and the offshore low-level "jet"). Atlanta, GA: Presentation given at the American Meteorological Society; 2010.
5. Eecen PJ, Branlard E. The OWEZ meteorological mast—analysis of mast-top displacements. Netherlands: Energy Research Centre of the Netherlands; 2008.
6. Zhang H-M, Bates JJ, Reynolds RW. Assessment of composite global sampling: sea surface wind speed. Geophys Res Lett 2006;33:L17714.
7. Beaucage P, Bernier M, Lafrance G, Choisnard J. Regional mapping of the offshore wind resource: towards a significant contribution from space-borne synthetic aperture radars. IEEE JSTARS 2008;1:48–56.
8. Jackson C, Apel J. Synthetic aperture radar marine user's manual. Washington, DC: U.S. Department of Commerce; 2004 p. 464. Available at http://www.sarusersmanual.com/. (Accessed 2012).

9. Liu WT, Tang W, Xie X. Wind power distribution over the ocean. Geophys Res Lett 2008;35:L13808.

SUGGESTIONS FOR FURTHER READING

Melnyk M, Andersen R. Offshore power: building renewable energy projects in U.S. waters. PennWell Corp; 2009. p. 496.

Peinke J, Schaumann P, Barth S, editors. Wind energy: proceedings of the Euromech colloquium. USA: Springer; 2011. p. 363.

Twidell J, Gaudiosi G, editors. Offshore wind power. UK: Multi-Science Publishing; 2009. p. 425.

15

UNCERTAINTY IN WIND RESOURCE ASSESSMENT

Wind resource estimates are only useful if their uncertainty is well defined. Unless the analyst can offer a degree of confidence that the resource falls within a specified range, it is not possible to construct a sound financial model for a wind project investment. Financial models depend on risk, and for a wind project, risk depends strongly on uncertainty in the resource.

The uncertainty present in all wind resource estimates is primarily related to the following factors: wind speed measurements, their relationship to the historical climate, potential future climate deviations, wind shear, and the spatial wind resource distribution. This chapter reviews these factors and provides a range of estimates for each. In addition, ways of combining the uncertainties from different sources depending on their correlation with one another are discussed. (Note that except where otherwise stated, the uncertainty estimates mentioned here are expressed as a percent of the speed and represent one standard error of a normal distribution.)

Not addressed in this chapter is the relationship between the uncertainty in speed and the uncertainty in energy production, which varies depending on the turbine model, mean wind speed, speed frequency distribution, and other factors. The uncertainty in turbine performance and losses is also not considered. These topics, along with other elements of energy production estimation, are discussed in the next chapter.

Wind Resource Assessment: A Practical Guide to Developing a Wind Project, First Edition.
Michael Brower et al.
© 2012 John Wiley & Sons, Inc. Published 2012 by John Wiley & Sons, Inc.

15.1 MEASUREMENT UNCERTAINTY

This is the uncertainty in the free-stream wind speed as measured by the anemometers after data validation and adjustments. It reflects not just the uncertainty in the sensitivity of the instruments when operating under ideal wind-tunnel conditions but also their performance in the field, where they may be subject to turbulent and off-horizontal winds, the possible effects of the tower and other obstacles on the observed speeds, and problems such as icing that may be missed in the validation. There may be additional uncertainties associated with certain anemometers and anemometer types, including those resulting from manufacturing or design flaws or damage incurred in the field.[1]

The uncertainty associated with anemometer response under ideal conditions (called the *sensor response uncertainty*) is typically estimated to be from 1.0% to 1.5% for a single anemometer. Considering that they are used for power curve testing and thus represent the de facto standard for estimating turbine output, class I sensors may be assumed to have a somewhat lower uncertainty than other sensors. The other components of the measurement accuracy vary depending on the circumstances. High turbulence, significant vertical winds, short booms or other factors contributing to tower effects on the free-stream speed, can all lead to greater uncertainty. The general range of estimates for a single anemometer mounted in accordance with the guidelines presented in this book, assuming good data quality, is around 1.5–2.5%.

The measurement error can be reduced by averaging the data from two sensors mounted in different directions at the same height on the mast. For those direction sectors where neither sensor is in the direct shadow of the tower, it is possible to reduce the measurement uncertainty by a factor of up to the square root of two (1.414), implying a combined uncertainty range as low as 1.1–1.8%. This strategy also reduces the risk of systematic error introduced by tower effects. For instance, with two anemometers mounted at right angles to each other, when one is directly upwind of the tower, the other points to the side. The first sees a decrease, and the other an increase, in the free-stream speed. As a result, the average of the two measurements is nearly unbiased.

The benefit of averaging is lessened when there are significant biases affecting both sensors, however. Possible examples include the effects of turbulence and vertical winds, which are likely to be similar for both sensors if they are the same model. Where it is suspected these effects might be substantial, only partial credit for the averaging should be taken.

15.2 HISTORICAL WIND RESOURCE

This uncertainty addresses how well the site data (after any MCP adjustments) may represent the historical norm. It is related to the amount of on-site data, the interannual

---

[1] A significant real-world example of a manufacturing or design flaw is the problem of "dry friction whip," a vibratory mode experienced by a portion of NRG Maximum 40 anemometers manufactured between May 2006 and December 2008. The problem typically reduces average wind speeds by up to several percent and causes unusually wide scatter compared to a problem-free reference anemometer (1, 2).

variability of the wind climate, the length of the historical reference period, and the correlation of the site data with the long-term reference. Referring again to the equation from Chapter 12,

$$\sigma_{\text{historical}} \cong \sqrt{\frac{r^2}{N_R}\sigma_R^2 + \frac{1-r^2}{N_T}\sigma_T^2} \tag{15.1}$$

Here, σ_R and σ_T are the standard deviations of the annual mean wind speeds (a measure of the interannual variability of the wind climate) for the reference (R) and target (T) sites, respectively, over a suitably long period; N_R and N_T are the number of years of reference and overlapping reference–target data, respectively; r is the Pearson correlation coefficient based on a suitable averaging period (such as daily means); and $\sigma_{\text{historical}}$ is the uncertainty in the derived historical mean wind speed at the target site.

As discussed in Chapter 12, this equation makes a number of assumptions, the most important of which is that the reference data record has been consistent through time, with no discontinuities or trends resulting from changing location, equipment, surroundings, and other factors. Given the difficulty of confidently meeting these requirements with reference data older than about 10–15 years, it is best not to allow N_R to exceed 15. In addition, because of seasonal effects, the equation should not be used with much less than 1 year of overlapping target and reference data.

Even when these conditions are met, actual errors may depart substantially from the equation. Figure 15-1 shows the results of an experiment to test the uncertainty of

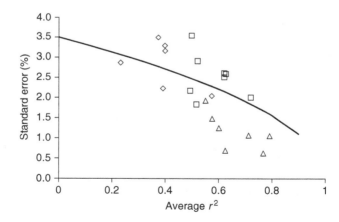

Figure 15-1. The results of an experiment to determine the uncertainty of MCP for three tall towers in different parts of the United States. Each point represents the standard error of the predicted mean speeds based on 12-month unconstrained linear regressions between a surface reference station and a tower. The smooth curve is the theoretical uncertainty based on Equation 15.1, assuming a 3.5% interannual standard deviation, the average for the three towers. *Source:* Taylor M, et al. An analysis of wind resource uncertainty in energy production estimates. AWS Truepower; 2004.

MCP using data from three tall towers. While the results for two of the three towers (diamonds and squares) are generally consistent with the theoretical curve based on Equation 15.1, the errors for the third tower (open triangles) are considerably smaller than predicted.

Historical wind records suggest that the standard deviation of annual mean wind speeds typically falls between 3% and 6% depending on the location and data source. In some regions, the variation may be either larger or smaller. Absent a thorough analysis of historical wind data for a particular region (something that is not always possible given problems with the long-term consistency of wind measurements), it is reasonable to assume a value of 4%. For 1 year of overlapping data, a reference record ranging from 7 to 15 years, and a correlation factor ranging from 0.6 to 0.9, the result of Equation 15.1 is an uncertainty range of 1.6–2.8%. Where suitable reference stations are lacking, the uncertainty is simply that of the period of on-site observation (for example, for 1 year, 4%).

15.3 FUTURE WIND RESOURCE

The uncertainty in the future wind resource can be divided into two components: that due to normal variability of the wind climate and that due to the risk of long-term climate change. Assuming the two components are unrelated, the individual uncertainties can be combined by taking the square root of the sum of the squares,[2] as follows:

$$\sigma_{future} = \sqrt{\sigma_{normal}^2 + \sigma_{climate}^2} \qquad (15.2)$$

For the first component, σ_{normal}, the same interannual variability used for estimating the historical climate uncertainty can be assumed. Adapting Equation 15.1, we have,

$$\sigma_{normal} \cong \frac{\sigma_R}{\sqrt{N_p}} \qquad (15.3)$$

where N_p is the number of years over which the average is to be calculated. This could be the financial horizon of the wind project investment, which typically ranges from 10 to 25 years. Assuming an interannual variation of 4%, the normal component of the uncertainty is 1.3% over 10 years and 0.8% over 25 years.

Although the climate-change component of the uncertainty is more speculative, it should not be ignored. Considering the studies conducted to date (reviewed in Chapter 12), a plausible range of uncertainty due to climate change is 0.5–2%. The

---

[2]The sum of the squares rule for independent sources of error can be derived from the equation for the variance (standard deviation squared) of a sample of data. When two large samples of data are combined, the variance of the combined distribution contains linear cross terms that tend to cancel out because they are randomly distributed about the mean, leaving only the squared terms. The result is that the variance of the combined distribution equals the sum of the variances of the individual distributions. For more information, see any standard text on statistics.

lower end of this range applies to a horizon of 10 years, while the upper end applies to a horizon as long as 25 years.

Combining the two components leads to an uncertainty of 1.4% for a 10-year project horizon and 2.2% for a 25-year project horizon. The climate-change component is negligible in the first instance—leaving it out would reduce the uncertainty by just 0.1%. It becomes increasingly important as the horizon increases.

It should be noted that these uncertainty estimates ignore the time value of money. In a present-value analysis supporting a plant investment decision, plant production, and hence revenues, in the distant future would be discounted more heavily than plant production in the near future. Taking this into account would tend to moderate the perceived financial risk associated with climate change, but would increase that associated with normal climate fluctuations.

15.4 WIND SHEAR

The wind shear uncertainty can likewise be divided into two components: the uncertainty in the observed wind shear due to possible measurement errors and the uncertainty in the change in wind shear above mast height. The two components are independent of one another, so the sum-of-the-squares rule applies.

The first component can be estimated using the following equation, which we first saw in a slightly different form in Chapter 10:[3]

$$\Delta\alpha_{obs,v} \cong \frac{\log(1 + \sigma_{r,v})}{\log\left(h_2/h_1\right)} \tag{15.4}$$

The numerator contains the uncertainty in the speed ratio, $\sigma_{r,v}$. Under most circumstances, this is approximately the uncertainty in the measured speed at each height multiplied by the square root of two.[4] For a speed ratio uncertainty ranging from 1.4% to 3.5% (corresponding to an uncertainty for a single anemometer ranging from 1.0% to 2.5%), and for upper and lower heights (h_2 and h_1) of 60 and 40 m, respectively, the uncertainty in the shear exponent $\Delta\alpha$ works out to be 0.034–0.085.

Note that any differences in the influence of the tower or other equipment on the speeds measured by the two anemometers could increase this uncertainty (assuming no correction can be made). Examples include when the uppermost sensors are placed too close to the top of the tower, when the sensor booms are not pointing in the same direction, and when the ratio of boom length to tower width varies.

---

[3]Unlike other uncertainties quoted in this chapter, the wind shear uncertainty is expressed as a deviation in the magnitude rather than as a proportion of the mean value. We use the delta prefix (Δ) to remind the reader of this distinction. One reason for this convention is that shear exponents can take on small or even negative values, making a percentage uncertainty difficult to interpret. In addition, a deviation in shear exponent translates directly into a percentage deviation in the hub-height speed (Eq 15.6).

[4]This follows because errors in the speed measurements at each height are assumed to be uncorrelated. Thus, the uncertainty in their ratio equals the square root of the sum of the squares of the uncertainties in each speed. Assuming the two speed uncertainties are the same, the result is the square root of two times the speed uncertainty.

Equation 15.4 ignores any effect of uncertainty in the heights of the instruments. The contribution of the height uncertainty can be estimated from the following equation:

$$\Delta\alpha_{obs,h} \cong \alpha \frac{\log(1 + \sigma_{r,h})}{\log\left(h_2/h_1\right)} \tag{15.5}$$

This is very similar to Equation 15.4, except that the speed ratio uncertainty is replaced by the height ratio uncertainty, $\sigma_{r,h}$, and there is an additional factor of the shear exponent α. Since heights can be determined quite accurately in the field, and since α is almost always much less than 1, this component of the uncertainty is usually much smaller than that associated with the speed ratio. Nevertheless, its contribution can be significant in some instances (such as when it is not possible to visit the tower to verify the instrument heights), and where that is so, it should be added to the speed-related uncertainty through the sum of the squares.

The second component, the uncertainty in the change in wind shear above mast height, is more difficult to estimate and depends very much on the site. As a rule of thumb, based on data from very tall towers, AWS Truepower estimates the uncertainty in the shear exponent above the top of the mast to be 10–20% of the observed shear depending on the complexity of the terrain and land cover. If the observed shear is 0.20 and the terrain is flat and open, an uncertainty of 0.02 might be assumed; if the observed shear is 0.30 and the terrain is complex, the uncertainty could be as high as 0.06. (Note that in some cases it is possible to reduce the uncertainty through the use of remote sensing.)

Combining the two components produces an overall uncertainty in the shear exponent ranging from about 0.04 to 0.10. The corresponding uncertainty in the hub-height speed is approximated by the following equation:

$$\sigma_{v,hh} = 100\left[\left(\frac{h_h}{h_2}\right)^{\Delta\alpha} - 1\right](\%) \tag{15.6}$$

The result is sensitive to the relative heights as well as, of course, the shear uncertainty. For a mast height of 60 m and hub height of 80 m, the range of uncertainty in the hub-height speed is 1.1–3.0%. For a mast height of 50 m and hub height of 90 m, the upper bound increases to 6.3%.

15.5 WIND FLOW MODELING UNCERTAINTY

The wind flow modeling uncertainty[5] can be defined as the uncertainty in the average wind speed at any point relative to the observed wind speeds at the site masts. The

---

[5]This chapter focuses on the uncertainty associated with numerical wind flow models; however, the same methods can be applied to other quantitative models as well.

range of uncertainty in practice can be very wide—from as low as 2% in simple, open terrain to 10% or more in complex terrain—and depends on the model used, the model's resolution, the terrain and wind climate, the placement of the masts, and other factors.

Ideally, the uncertainty is determined directly from the on-site wind measurements. This is possible when the following conditions are met:

- There are at least 5 and preferably 10 or more masts in the project area.
- The masts are well distributed within the proposed turbine array and among the wind conditions likely to be experienced by the wind turbines.
- There is sufficient data from each mast to accurately compare mean annual wind speeds.

Here is the procedure. First, one of the masts is designated the reference mast. Next, the mean wind speed at each of the other masts is predicted from the reference mast using the wind flow model. The error between the predicted and observed mean wind speeds is then calculated. The process is repeated with each of the other masts serving as the reference. Finally, the standard deviation of all the errors is calculated to estimate the wind flow modeling uncertainty.

More often than not, the conditions listed above are not met. In that case, the resource analyst must rely on experience (preferably with the same model) at similar sites. Proceedings of technical meetings and conferences are a useful reference for this purpose. An assessment of the complexity of the terrain, variations in land cover, placement of the masts, and the possible role of coastal sea breezes and other atmospheric circulations all come into play here.

15.6 COMBINING UNCERTAINTIES

Once the various components of the uncertainty have been defined, they must be combined in some way to arrive at the uncertainty in the average wind speed of the entire turbine array. Before this can be done, it is necessary to consider whether the various sources of error are *correlated* or *uncorrelated*. Uncorrelated errors are independent of one another—the sign and magnitude of one error has no bearing on the sign and magnitude of another. Imagine a fleet of small boats floating on a choppy sea. Each boat rides its own wave, its motion unrelated to the motions of the other boats. Correlated errors, on the other hand, march in lockstep with each other. The corresponding analogy is a tide that lifts or lowers all the boats at once.

It is better for uncertainties to be uncorrelated because then they do not add in a linear fashion but rather as the sum of the squares. This reduces the combined uncertainty. For example, suppose the measurement uncertainty and the shear uncertainty are both 2%, and uncorrelated. The combined uncertainty is

$$\sqrt{0.02^2 + 0.02^2} = 2.8\%$$

If, on the other hand, the uncertainties are perfectly correlated, then the combined uncertainty is their linear sum, $0.02 + 0.02 = 4\%$, which is 40% larger.

The challenge is to determine which errors are correlated and which are not. (Some errors may be partially correlated; however, such cases can usually be dealt with by dividing the uncertainties into separate components that are either purely correlated or purely uncorrelated.) By convention, the different components of the uncertainty for a single mast are assumed to be uncorrelated and can therefore be combined as the sum of the squares. Although one can think of situations where this does not hold completely true,[6] it is generally a good assumption. The uncertainty ranges discussed earlier for the separate components can therefore be combined into an overall range for a single mast, as shown in Table 15-1.

When uncertainty estimates from different masts are combined, more attention must be paid to the possibility of correlated errors. Here are some examples:

- When the masts at a project site span the same or a similar period, their long-term mean wind speed estimates may be biased in a similar way.
- Any deviation of the future wind resource from the historical average will apply equally to all parts of the project, regardless of the number of masts used in the resource assessment.
- Wind flow modeling errors can be systematically biased if the masts are situated in one type of terrain (for example, along a ridgeline), while some of the turbines are in a different type of terrain (down the slope).
- Tower effects on speed measurements may introduce a consistent bias if the anemometers are all mounted in the same direction relative to the prevailing wind (of particular concern when there is only one anemometer mounted at each height).

Table 15-1. Summary of general uncertainty ranges by category for a single mast, equipped, maintained, and analyzed according to the guidelines described in this book. For an explanation of each uncertainty range, see the corresponding text.

Category	Uncertainty, %
Measurement accuracy (single anemometer)	1.0–2.5
Historical wind resource	1.6–4.0
Future wind resource (plant life of 10 or 25 yr)	1.4–2.2
Wind shear	0.0–6.3
Wind flow modeling	2.0–10.0
Total uncertainty	**3–13**

It is assumed that the various components are uncorrelated with one another, and thus the total uncertainty is the square root of the sum of the squares of the individual uncertainties.
Source: AWS Truepower.

[6]One such case was mentioned earlier: turbulence or vertical winds may bias all anemometers in a similar way. When this happens, averaging the wind speeds from two anemometers at the same height does not reduce the uncertainty. As another example, when the uppermost anemometers are placed too close to the tower top, there may be a tendency to overestimate both the mean speed and the shear.

- At some sites, the shear may change with height above the masts in a consistent way, resulting in a systematic bias in the projected hub-height wind speed.

Each component of the uncertainty for a mast should be combined with the same component for other masts in either a weighted linear sum or a squared sum depending on whether the component is correlated or uncorrelated. The weight given to each mast should be proportional to its influence on the overall array-average wind speed. One approach to determining the weights is to divide the array into groups of turbines, each associated with a particular nearby mast. This leads to the following equations:

Uncorrelated

$$\sigma_{combined} = \frac{\left(\sum_{i=1}^{M} N_i^2 \sigma_i^2\right)^{1/2}}{N_T} \tag{15.7}$$

Correlated

$$\sigma_{combined} = \frac{\sum_{i=1}^{M} N_i \sigma_i}{N_T} \tag{15.8}$$

Here, the sums are over M, the number of masts; N_i is the number of turbines associated with mast i; σ_i is the percent uncertainty in the average speed for that group of turbines; and N_T is the total number of turbines in the array. (Where some form of smooth blending of the predicted speeds from different masts is applied, the uncertainty equations naturally become more complicated, but the same general principles hold.)

Equation 15.7 implies that the uncorrelated portion of the uncertainty decreases the more evenly distributed the masts are among the turbines. The effect is demonstrated in Figure 15-2 for the case of two masts, each having an uncertainty of 10%. If one mast is the predominant influence for half the turbines and the other mast is the predominant influence for the other half, the combined uncertainty is reduced to 7.1%. In fact, assuming the individual uncertainties are the same for all masts (call it σ_0) and an equal number of turbines are associated with each mast, Equation 15.7 reduces to

$$\sigma_{combined} \cong \frac{\sigma_0}{\sqrt{M}} \tag{15.9}$$

This is the familiar equation describing how the uncertainty in the mean value of a quantity goes down with the square root of the number of independent measurements of that quantity. At the opposite extreme, if all the turbines are associated with just one mast i (e.g., because the other masts are much farther from the turbines), then

$$\sigma_{combined} = \sigma_i \tag{15.10}$$

These equations confirm that the uncertainty in the array-average wind speed is strongly affected not only by the number of wind monitoring masts but also by their placement within the proposed turbine array. They demonstrate clearly why—as we have stressed—it is important to place the masts in representative locations throughout the array.

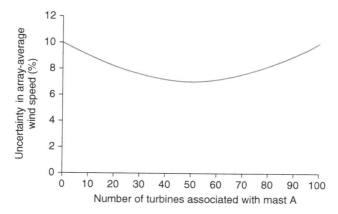

<u>Figure 15-2.</u> Variation of the combined uncorrelated uncertainty in the array-average wind speed for the case of two masts as a function of the number of turbines associated with one of the masts (mast A). The rest of the turbines are assumed to be associated with mast B. The uncertainty based on either mast alone is assumed to be 10%. *Source:* AWS Truepower.

Now, all the elements are in place to combine the uncertainties from multiple masts. Although the process may seem rather involved at first glance, Equations 15.7 and 15.8 can be applied easily in a spreadsheet program once the essential decision, whether a particular uncertainty component is to be treated as correlated or uncorrelated between masts, is made. Table 15-2 provides an example with five masts, for which it has been assumed that the historical and future wind resource uncertainties are correlated across masts (but not with each other) and that all other uncertainties are uncorrelated. To be done correctly, the uncertainties should *not* be totaled for each mast first. Instead, each separate uncertainty component should be combined across masts (i.e., down each column, as indicated by the arrows) according to whether it is correlated or uncorrelated. Then the combined uncertainties should be totaled (i.e., across the bottom row) assuming they are all uncorrelated.[7]

The resulting total combined uncertainty in this case, 4.3%, is part way between the results that would be obtained when assuming all uncertainties are uncorrelated (3.1%) and when assuming all the uncertainties are correlated (6.7%). This shows how purely correlated and uncorrelated errors combine to produce a partial correlation.

As a final note of caution, the assumption that uncertainties are uncorrelated is a tempting one, given the benefits that accrue from using the sum-of-the-squares rule; but it can be optimistic. The history of wind energy development is replete with cases in which it was assumed that errors from different sources were random and unrelated, whereas they were in fact biased in consistent ways. For example, at one time it was common practice to mount anemometers vertically off the tops of towers on short

---

[7]A more sophisticated uncertainty analysis can be performed using statistical packages available from a number of vendors such as Palisades, GoldSim, and Lumenaut. These packages are designed to simulate complex statistical problems involving numerous partially correlated and covariant variables.

Table 15-2. An example calculation of the array-average uncertainty for a layout of 65 turbines associated with five different masts. The arrows indicate the order of calculation. First, the columns of correlated uncertainties are combined according to Equation 15.8, and the columns of uncorrelated uncertainties according to Equation 15.7. Next, the total uncertainties in the bottom row are combined assuming the individual components are uncorrelated.

MAST	No.of Turbines (N)	Historical Resource (Correlated)	Future Resource (Correlated)	Other Uncertainties (Uncorrelated)	Total
A	15	3.0%	1.5%	5.5%	6.4%
B	7	3.0%	1.5%	6.2%	7.0%
C	20	3.0%	1.5%	5.8%	6.7%
D	9	3.0%	1.5%	6.0%	6.9%
E	14	3.0%	1.5%	5.6%	6.5%
Combined	65	3.0%	1.5%	2.7%	4.3%

Source: AWS Truepower.

stub masts. Because of the acceleration of wind over the tower top, this introduced a positive bias in both the observed wind speed and the wind shear, resulting in a doubly serious overestimation of the projected hub-height wind speed. Similarly, it used to be common practice to place towers at the very best—most exposed, highest elevation—points within a project area. This, too, led at times to biased wind resource estimates because of the tendency of wind flow models to underestimate the drop in wind speed with decreasing elevation.

Given this history, the conservative analyst might choose to assume that all the errors associated with one mast are correlated with those of other masts. This assumption, however, takes no credit at all for deploying multiple masts at a project area and could result in an overly pessimistic assessment of the financial risks in the wind project. By judging each source of error on its merits and treating them in a consistent way, it should be possible to arrive at an objective assessment of the wind resource uncertainty.

15.7 QUESTIONS FOR DISCUSSION

1. What is uncertainty? Why is it important to consider the uncertainty band in a given wind resource assessment analysis?

2. What are the primary factors that are typically considered when quantifying the uncertainty of a wind resource assessment program? What are some specific factors that determine the magnitude of each?

3. Why would averaging concurrent valid observations from two anemometers at the same level reduce the overall measurement uncertainty?

4. As discussed in Chapter 12, it is important to adjust the observed data record to represent the long-term climate conditions, where possible. Explain the impact of this adjustment from an uncertainty perspective.

5. If there is a trend in the data from a long-term reference station used for MCP, should its impact on the results be treated as an uncertainty? Why or why not? Using the Internet or other resources, research and discuss the differences between uncertainty and bias.

6. Assume your wind monitoring mast records the wind speed at heights less than your proposed turbine hub height. Referring to the equations in this chapter, what are some specific measures that can be taken in the design of the monitoring setup and the analysis of the data to reduce the uncertainty associated with the extrapolation of the observed values to hub height?

7. What are some ways to design a wind resource assessment campaign that would minimize the wind flow modeling uncertainty for a proposed project in complex terrain? Explain why these practices could be effective. What would you have to do to eliminate this uncertainty altogether?

8. Suppose the wind flow modeling uncertainty for any point within a project area, based on any single mast, is 8%. Assuming the masts are distributed evenly throughout the array, what would be the uncertainty in the array-average speed if there were one, four, and eight masts installed? Take the case of four masts. Suppose half the turbines were assigned to one mast and the other half were divided equally among the remaining three. What would be the array-average uncertainty then?

9. Assume each additional mast in the previous problem costs $25,000 to install and operate for 1 year. What is the incremental cost *per percentage point reduction in uncertainty* gained by going from one mast to four masts and from four to eight masts? Assume an even distribution of masts within the array.

10. Suppose your wind project has two masts. One mast is associated with the northern portion of the project area containing 20 turbines; the other is associated with the southern portion containing 30 turbines. You have estimated the uncertainty for each of the components and determined whether they are correlated or uncorrelated (see the table below). Using Equations 15.7 and 15.8 and the method outlined in Table 15-2, estimate the combined uncertainty in the array-average wind speed.

Source	Mast A, %	Mast B, %	Correlated/Uncorrelated
Measurement	1.5	2.0	Uncorrelated
MCP	3.0	3.0	Correlated
Future resource	2.2	2.2	Correlated
Shear	3.0	2.0	Correlated
Wind flow modeling	4.0	4.0	Uncorrelated

REFERENCES

1. Clark S., et al. Investigation of the NRG #40 anemometer slowdown. NRG Systems. Presented at Windpower, Chicago (IL); 2009.
2. Hale E, Fusina L, White E, Brower M. Correction factors for NRG #40 anemometers potentially affected by dry friction whip. USA: AWS Truepower, 2011.

SUGGESTIONS FOR FURTHER READING

Coquilla RV, Obermeier J, White BR. Calibration procedures and uncertainty in wind power anemometers. Wind Eng 2007;31(5):303–316.

Garson GD. Statnotes: topics in multivariate analysis. Retrieved http://faculty.chass.ncsu.edu/garson/pa765/statnote.htm. (Accessed 2012).

Lackner MA, Rogers AL, Manwell JF. Uncertainty analysis in MCP-based wind resource assessment and energy production estimation. J Solar Energy Eng 2008;130:pp. 1–10.

Wilks DS. Statistical methods in the atmospheric sciences. USA: Academic Press; 1995.

PLANT DESIGN AND ENERGY PRODUCTION ESTIMATION

The methods and challenges of designing a wind power plant and estimating its energy production could easily fill their own book. The purpose of this chapter is to round out the discussion of wind resource assessment by providing an overview of this subject. The reader is introduced to the underlying principles and methods of energy production estimation and taken through the main steps of selecting an appropriate turbine model, importing or creating a wind resource grid (WRG), wake modeling, designing and optimizing a layout, and estimating losses.

16.1 PLANT DESIGN SOFTWARE

One of the first steps in designing and evaluating a wind energy project is selecting the plant design software (Fig. 16-1). There are several software packages available on the market, including GL Garrad Hassan's WindFarmer, ReSoft's WindFarm, EMD International's WindPro, and AWS Truepower's openWind. These programs all have a number of features and capabilities in common. With them users can

Wind Resource Assessment: A Practical Guide to Developing a Wind Project, First Edition.
Michael Brower et al.
© 2012 John Wiley & Sons, Inc. Published 2012 by John Wiley & Sons, Inc.

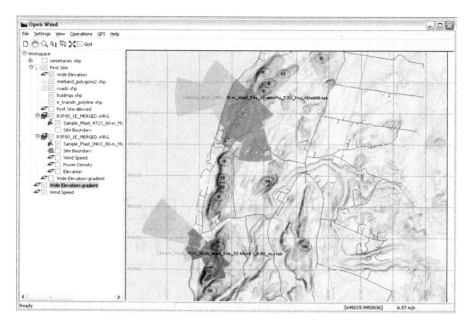

<u>Figure 16-1.</u> Screenshot of the main screen of the openWind wind power plant design and optimization software. *Source:* AWS Truepower.

- import wind frequency distributions from one or more monitoring towers;
- create or import the results of a wind flow simulation;
- characterize turbines according to a number of parameters including hub height, rotor diameter, and power output and thrust over a range of speeds;
- place the turbines within the project area;
- estimate gross energy production (without losses);
- calculate wake losses and apply other losses to arrive at the net energy production.

These features constitute the basic elements of an energy production simulation. Most software packages also have the ability to adjust the turbine positions automatically to maximize the net energy production while meeting certain constraints (such as respecting property boundaries and setbacks). This process is called *optimization*. It generally involves striking a balance between placing turbines close together in the windiest locations and spacing them far enough apart to keep wake losses (and the extra wear and tear caused by wake-induced turbulence) to a minimum. Although there are exceptions, this usually means, for projects on land, that turbines are spaced no closer than about 6–10 rotor diameters apart along the most frequent wind directions and 3–4 rotor diameters apart along the least frequent directions. Offshore, the spacing is usually larger because wakes tend to persist for a greater distance.

The programs also offer a variety of optional capabilities, which vary to some degree with the software. These capabilities permit the user to

- calculate noise levels and constrain the layout to meet noise limits at specified boundaries or points;
- map viewsheds (zones of visual impact) and constrain the layout to minimize turbine visibility from specified points;
- model turbine shutdown strategies designed to reduce wear and tear caused by wakes from nearby turbines (called *sector management* or *directional curtailment*);
- blend wind flow modeling from multiple monitoring masts;
- perform uncertainty analyses;
- design access road and electrical collection systems and estimate their costs.

Before choosing a program, the resource analyst should investigate its features, design limits (such as the maximum number of turbines that can be modeled), user interface, required operating system and computer platform, compatibility with other file types, and, of course, price and technical support options. Beginners should choose software that is relatively easy to use and has a strong support package. It is a good idea to try out a free demonstration copy (or, in the case of openWind, the free, open-source version) before purchasing.

16.2   SETTING UP THE PROJECT

The initial project setup usually involves importing into the software a variety of graphical image backdrops, raster layers, and vector layers describing the important geographical and geophysical features of the project area. Among the features to consider at the outset are the following:

- terrain elevations;
- land cover types;
- rivers and water bodies;
- roads and trails;
- transmission and distribution lines;
- administrative boundaries (towns, counties, provinces, and so forth);
- property and other land parcel boundaries;
- locations of residences;
- parks, military reservations, and other potentially excluded areas;
- locations of large structures such as buildings and communications towers;
- locations of significant cultural, religious, touristic, or other landmarks that may affect siting decisions.

These data can be obtained from a variety of sources, some of which are listed in Appendix B. Data layers can also be created through the digitizing of aerial photographs or topographic maps. (This may require some GIS expertise and digitizing equipment.)

It will be useful for the analyst to be familiar with the three types of geographical and geophysical files that are most often used and with their main purposes.

- *Backdrop images* are for visual reference only. They help orient the user on the landscape and make sure all other features are properly aligned (an example is shown in Figure 16-2). Backdrops can include aerial photographs and topographical maps, which are often available in a georeferenced format such as geotiff, or in a standard graphical format (bmp, tif, jpeg) accompanied by a georeferencing file such as an ESRI ArcInfo "world" file. If no such georeferencing information is available, then it may be necessary to create it within a GIS.

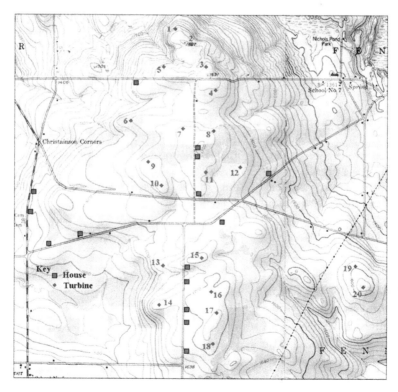

Figure 16-2. Backdrop image from a USGS topographic map, overlaid with point layers showing houses (red squares) and turbines (blue diamonds) and polygon layers showing the project boundaries. *Source:* AWS Truepower.

- *Raster layers* consist of data values on an evenly spaced grid, usually with one value for each grid cell or node (though interleaved and other multidimensional formats exist). The files can contain digital terrain models (DTMs) or digital elevation models (DEMs), land cover classes, surface roughness values, wind speeds, definitions of allowed or excluded development areas, and many other types of information. Aside from providing a visual reference, such information can be used directly in the analysis. For example, if the software includes a wind flow model, it can make use of the elevation model and roughness map in the simulation. Elevation models can be used directly or to derive other information, such as slopes, to determine where turbines can be placed. Land cover classification maps can be used to define areas, such as water bodies or wetlands, excluded from development.
- *Vector layers* are made up of points, lines, or polygons. They are most often used to define locations and extents of important features such as buildings and water bodies, as well as the boundaries of administrative areas (e.g., cities and counties), land parcels, and the project area itself. Another common use is for defining topographic contours, which represent lines of constant elevation. Although almost anything that can be communicated in a vector layer can be transformed into a raster layer and vice versa, vector layers are not limited by spatial resolution in the same way that raster layers, with their fixed grid spacing, are. This makes them well suited for defining boundaries and locations very precisely. They can also contain much additional information in associated descriptive database files, such as the names and addresses of landowners, the names of water bodies, or the voltages of transmission lines.

During the project setup, the user must define the boundaries of where turbines can be placed. This is usually the land that is controlled by the developer or by others who have reached an agreement to lease the land to the developer. One way this could be accomplished is by selecting parcels of land already defined in a vector layer. From the selected parcels, a new vector layer can be created. Another way is to draw the boundaries manually. Depending on the software, these methods may require a GIS.

16.3 WIND RESOURCE DATA

Once the project has been set up, the next step is usually to import or create the wind resource information required for the simulation. This is where everything covered in this book until now comes into play. Two main types of wind resource information are used in wind plant design software: measurements from a point location such as a wind-monitoring mast, and the output of a numerical wind flow model describing the variation of the wind resource across the project area. In most cases, both types of information are required for an accurate simulation, and they are used together.

16.3.1 Wind Resource Measurements

Wind resource measurements are the linchpin of energy production simulations. Only direct measurements from within the project area capture with high confidence the true speed and direction frequency distributions that determine the energy production potential. Without them, it is rare that financial institutions or investors will risk money on building a wind project of a significant size.

The wind resource data are usually imported into wind plant design software in the form of a tabular file popularly called a *TAB file*. One such file is imported for each mast in the project area to be used in the energy production simulations. A TAB file provides the frequency of occurrence of winds observed within specified speed and direction bins. An example of this format is shown in Figure 16-3. In principle, a time series of wind measurements could also be used, but only some programs support this option.

```
Untitled | Norm=  49991.0  True Vmean=   8.11  PMean=   738.6
   545122        4731825   80.0
       12   1.00      0.00
           6.035    2.941    2.825    3.135    9.452   17.790    8.210    2.381    2.106    3.890   18.805   22.429
 1.00     32.12    58.77    60.48    48.43    16.81    11.57    23.73    71.33    82.86    43.93    11.32     9.71
 2.00     94.30   145.88   176.65   156.63    44.41    27.18    57.50   134.41   160.67    74.32    19.31    26.89
 3.00    144.75   210.82   217.36   178.64    76.63    43.41    71.14   146.59   125.84    93.86    26.11    38.20
 4.00    183.23   196.54   174.17   174.49   101.50    67.29    94.05   119.71    87.85    57.14    28.14    57.49
 5.00    162.09   120.71   119.65   115.16   111.11    80.57   114.75    99.54    91.18    71.43    34.46    75.45
 6.00    112.03   108.61    67.26   104.31   106.35    81.07   101.53    94.08    83.10    98.74    45.42    73.96
 7.00     89.98    71.61    59.83    57.42   100.01    80.43    99.73    70.56    47.01    86.14    46.59    80.41
 8.00     47.58    34.00    44.60    58.70    93.99    79.25    84.29    68.88    57.93   102.51    51.96    79.73
 9.00     41.76    22.44    28.32    38.60    85.56    85.70    82.23    58.80    67.91    90.35    53.19    81.53
10.00     27.84    20.06    19.47    20.10    71.45    82.08    80.64    57.12    53.66    78.69    63.07    76.71
11.00     23.04     7.14    13.45    17.86    62.22    81.96    57.18    42.04    53.19    67.63    72.65    75.63
12.00     18.40     3.40     8.85    13.40    44.45    73.50    40.88    17.19    34.67    39.86    72.00    66.91
13.00     12.10     0.00     4.25     9.57    33.65    63.19    37.28    10.92    15.20    25.97    77.80    57.51
14.00      4.14     0.00     3.54     3.83    24.23    41.27    25.09     4.62    10.45    21.34    78.37    46.02
15.00      2.98     0.00     1.42     1.59    15.66    36.66    16.93     0.84     4.75    11.57    72.21    40.44
16.00      1.99     0.00     0.71     0.00     5.50    28.34     7.07     2.52     5.70    12.09    55.86    30.38
17.00      0.33     0.00     0.00     0.00     1.80    15.24     3.17     0.00     1.90     6.17    47.72    20.20
18.00      0.66     0.00     0.00     0.00     1.59    11.30     0.97     0.00     2.37     6.69    33.44    13.65
19.00      0.66     0.00     0.00     0.00     1.06     6.35     0.61     0.84     8.49     3.60    22.03    12.35
20.00      0.00     0.00     0.00     0.00     1.38     2.64     0.00     0.00     3.86     3.34    20.58    10.21
21.00      0.00     0.00     0.00     0.64     0.42     0.56     0.49     0.00     0.47     2.06    13.46     6.34
22.00      0.00     0.00     0.00     0.00     0.21     0.34     0.00     0.00     0.00     1.03    13.08     4.37
23.00      0.00     0.00     0.00     0.64     0.00     0.11     0.49     0.00     0.00     1.54    11.49     3.12
24.00      0.00     0.00     0.00     0.00     0.00     0.00     0.00     0.00     0.00     0.00     6.49     2.94
25.00      0.00     0.00     0.00     0.00     0.00     0.00     0.24     0.00     0.95     0.00     5.43     2.76
26.00      0.00     0.00     0.00     0.00     0.00     0.00     0.00     0.00     0.00     0.00     4.79     2.23
27.00      0.00     0.00     0.00     0.00     0.00     0.00     0.00     0.00     0.00     0.00     6.44     1.56
28.00      0.00     0.00     0.00     0.00     0.00     0.00     0.00     0.00     0.00     0.00     3.99     1.38
29.00      0.00     0.00     0.00     0.00     0.00     0.00     0.00     0.00     0.00     0.00     1.70     1.16
30.00      0.00     0.00     0.00     0.00     0.00     0.00     0.00     0.00     0.00     0.00     0.69     0.58
31.00      0.00     0.00     0.00     0.00     0.00     0.00     0.00     0.00     0.00     0.00     0.11     0.09
32.00      0.00     0.00     0.00     0.00     0.00     0.00     0.00     0.00     0.00     0.00     0.11     0.09
33.00      0.00     0.00     0.00     0.00     0.00     0.00     0.00     0.00     0.00     0.00     0.00     0.00
34.00      0.00     0.00     0.00     0.00     0.00     0.00     0.00     0.00     0.00     0.00     0.00     0.00
35.00      0.00     0.00     0.00     0.00     0.00     0.00     0.00     0.00     0.00     0.00     0.00     0.00
```

Figure 16-3. An example of a TAB file containing wind measurements binned by direction and speed. The first two lines contain information about the site, including, in the second line, the X and Y coordinates and height of measurement. The third line indicates the number of direction bins or sectors (in this case 12, indicating 30°-wide sectors), as well as scale and shift factors used for adjusting the speeds and directions, if necessary. The fourth line indicates the frequency of occurrence of each direction sector (e.g., in this case, winds come from the third sector, east–northeast, 2.825% of the time). At the start of each successive line is the upper limit of each wind speed bin (1 m/s in the first line), followed by the relative number of occurrences (summing to 1000 in this example) of that speed range within each direction sector. For example, when the wind direction is in the third sector, the speed falls in the 4–5 m/s speed bin with a frequency of 119.65 out of 1000, or 11.965%. The total frequency of occurrence in this bin is therefore 2.825% × 11.965% = 0.338%. *Source:* AWS Truepower.

The data should normally be entered for the hub height of the turbine to be modeled. While most software allows the data to be projected from a different height to the hub height, this means that any dependence of the wind shear on time of day and direction and other factors is lost. For reasons discussed in Chapter 11, this can produce errors in the frequency distribution of wind speeds at hub height and therefore in the estimated energy production.

It is very important to enter the mast coordinates as precisely as possible and to verify the position against the topographic data and landmarks within the software. Along with the modeled WRG (described in the next section), the model uses the mast position to extrapolate the observed wind resource to the turbines. In complex terrain, even a small error in the position can produce large errors in the estimated energy production.

16.3.2   Modeled Wind Resource Grids

The second key piece of resource information required by wind plant design software is a WRG. This is almost always generated by a numerical wind flow model of some kind. The conventional WRG format (made nearly universal by the WAsP software) is a text file containing predicted frequencies and Weibull parameters (A and k) for each of the 12 or 16 direction sectors for every point within the project area. An example is shown in Figure 16-4. The points are typically spaced anywhere from 10 to 50 m apart in a regular grid, providing good definition of the spatial variation of the wind resource.

A related file that is usually required is called a *point WRG*. This is the modeled wind resource at the precise location of each mast, in a format very similar to a WRG file but with a single line of data. Since masts are not generally found at exactly the center of a grid point, the point WRG values usually differ from those of the nearest grid point in the WRG. The information in the point WRG is used by the plant design software to calculate the speedup ratios for the extrapolation of the observed resource to other points in the project area. (The speedup ratio method is described in Section 16.7.1.)

Although the conventional WRG format is widely used, it has significant limitations, including limited data precision and an inability to incorporate wind resource data other than the Weibull factors (which, as discussed in Chapter 10, do not always represent actual wind speed frequency distributions with good accuracy). Improvements in this format are consequently desirable. One new file format that emerged recently is the plant data grid (PDG), which is supported by the openWind software and some wind flow models. This flexible, binary format has full floating precision and can support multiple data layers including turbulence, inflow angle, mean shear, and others.

16.4   SELECTING A TURBINE

Once the project is set up and the wind resource information has been loaded, the analyst is ready to design the turbine layout. This process starts by selecting a turbine model. Sometimes the choice is obvious; perhaps, the developer has a relationship

```
504      407   0.52687500E+06    0.39108250E+07    0.50000000E+02
 526875  3910825.  1060.8  80.0  9.1  2.09  0.7353E+03  12  66  61  228  31  35  179  29  35  176  33  38  193
 526925  3910825.  1057.4  80.0  9.0  2.09  0.7318E+03  12  66  60  228  31  34  179  29  35  176  33  38  192
 526975  3910825.  1053.7  80.0  9.0  2.09  0.7138E+03  12  66  60  227  31  34  180  29  35  176  33  38  192
 527025  3910825.  1050.2  80.0  9.0  2.09  0.7308E+03  12  66  60  226  31  34  180  29  35  176  33  38  191
 527075  3910825.  1047.5  80.0  9.0  2.09  0.7078E+03  12  66  60  225  31  34  180  29  35  175  33  38  189
 527125  3910825.  1044.1  80.0  9.0  2.09  0.7140E+03  12  66  60  224  31  34  181  29  35  176  33  38  188
 527175  3910825.  1042.4  80.0  8.9  2.09  0.6817E+03  12  66  60  223  31  34  181  29  35  177  33  37  187
 527225  3910825.  1041.4  80.0  8.9  2.10  0.6867E+03  12  66  60  222  31  34  182  29  35  177  33  37  187
 527275  3910825.  1040.0  80.0  8.9  2.10  0.6818E+03  12  66  59  221  31  34  182  29  35  177  33  37  186
 527325  3910825.  1038.3  80.0  8.9  2.10  0.6765E+03  12  66  59  220  31  34  183  29  35  177  33  37  185
 527375  3910825.  1036.5  80.0  8.9  2.10  0.6856E+03  12  66  59  219  31  34  184  29  35  177  33  37  184
 527425  3910825.  1035.7  80.0  8.9  2.10  0.6952E+03  12  66  58  218  31  34  184  29  35  177  33  37  184
 527475  3910825.  1034.7  80.0  9.0  2.10  0.6846E+03  12  66  58  217  31  35  185  29  35  177  33  37  183
 527525  3910825.  1033.7  80.0  9.0  2.10  0.6854E+03  12  66  58  217  31  35  187  29  35  178  33  37  182
 527575  3910825.  1032.3  80.0  8.9  2.10  0.6765E+03  12  66  58  217  31  35  187  29  35  178  33  37  182
 527625  3910825.  1031.1  80.0  9.0  2.10  0.6759E+03  12  66  58  216  31  35  188  29  35  178  33  37  181
 527675  3910825.  1030.7  80.0  9.0  2.10  0.6813E+03  12  66  58  216  31  34  188  29  35  178  33  37  181
 527725  3910825.  1031.9  80.0  9.0  2.10  0.6881E+03  12  66  58  215  31  35  188  29  35  178  33  37  181
 527775  3910825.  1032.6  80.0  9.0  2.10  0.6910E+03  12  66  58  215  31  35  189  29  35  178  33  37  180
 527825  3910825.  1033.4  80.0  8.9  2.10  0.6725E+03  12  66  58  215  31  35  189  29  35  178  33  38  180
 527875  3910825.  1034.1  80.0  8.9  2.10  0.6655E+03  12  66  57  215  31  35  189  29  35  178  33  38  180
 527925  3910825.  1035.0  80.0  8.9  2.10  0.6495E+03  12  66  57  215  31  36  189  29  36  178  33  38  180
 527975  3910825.  1034.9  80.0  8.8  2.10  0.6348E+03  12  66  58  215  31  36  190  29  36  179  33  38  180
 528025  3910825.  1035.8  80.0  8.8  2.10  0.6267E+03  12  66  58  215  31  37  190  29  36  179  33  38  180
 528075  3910825.  1036.4  80.0  8.9  2.10  0.6459E+03  12  66  59  215  31  37  190  29  37  179  33  38  180
 528125  3910825.  1035.6  80.0  8.8  2.10  0.6422E+03  12  66  60  215  31  38  190  29  37  179  33  38  180
 528175  3910825.  1040.4  80.0  8.9  2.10  0.6496E+03  12  66  60  216  31  38  190  29  37  179  33  38  180
 528225  3910825.  1047.5  80.0  9.0  2.10  0.6661E+03  12  66  60  216  31  39  191  29  38  180  33  39  180
 528275  3910825.  1046.6  80.0  9.0  2.11  0.6769E+03  12  66  60  214  31  39  193  29  38  180  33  39  178
 528325  3910825.  1039.7  80.0  9.0  2.11  0.6730E+03  12  66  60  214  31  39  193  29  38  181  33  39  178
 528375  3910825.  1035.3  80.0  9.0  2.10  0.6523E+03  12  66  60  215  31  39  193  29  38  181  33  39  179
 528425  3910825.  1033.1  80.0  8.9  2.10  0.6483E+03  12  66  58  216  31  38  194  29  37  181  33  38  179
```

Figure 16-4. An example of a WRG file containing the gridded output of a wind flow simulation. The first line indicates the number of columns and rows in the grid, the X and Y coordinates of the upper-left grid point, and the grid spacing. (Here scientific notation is used, but it is not required.) The next lines contain the wind resource data for each grid point. The first two parameters in each line are the X and Y coordinates of the point. These are followed by elevation above mean sea level (1060.8 m for the first grid point), the height above ground (80 m), the overall mean Weibull A parameter (9.1 m/s), the overall Weibull k parameter (2.09), and the mean wind power density (735.3 W/m^2). Then there is the number of direction sectors (12), and for each direction sector, in groups of three, the frequency (in percentage), the Weibull A (multiplied by 10), and the Weibull k (multiplied by 100). Note: for this example, the table is truncated to the right after the fourth direction sector. *Source:* AWS Truepower.

Table 16-1. IEC classifications for turbine suitability for the standard sea-level air density of 1.225 kg/m^3.

Wind Turbine Class	I			II			III		
	A	B	C	A	B	C	A	B	C
V_{ref} (m/s)	50	50	50	42.5	42.5	42.5	37.5	37.5	37.5
I_{ref}	16%	14%	12%	16%	14%	12%	16%	14%	12%

Source: IEC 61400-1 Third Edition 2005-8.

with a turbine vendor or has already placed a number of turbines on order for this and other projects. Very often, though, the developer is free to consider any of an ever-growing selection of wind turbine models.

However, not every turbine is suitable for every site. The question of turbine suitability should be addressed provisionally early in the process to avoid wasting time assessing unsuitable models or developing unrealistic expectations of the performance of the project. (The final decision on turbine suitability is up to the manufacturer, who must decide whether the turbine will be warranted for the site, and if so, under what operating conditions.)

The analyst should become familiar with the standard IEC turbine classifications, which are summarized in Table 16-1. It is common to hear sites referred to as "Class

IIB" or "Class IIIA" depending on their wind resource. In this table, V_{ref} is the largest value of the 10-minute average wind speed expected at hub height over a 50-year period, and I_{ref} is the expected average turbulence intensity at a hub-height speed of 15 m/s. Both can be estimated from on-site measurements. V_{ref} is defined for a standard sea-level air density of 1.225 kg/m3. While it may be possible to adjust it for sites with a much different air density, the manufacturer should be consulted to determine whether and how this should be done.

Class I turbines are designed to withstand the greatest wind loads. They are often, though not always, designed with smaller rotor diameters relative to their nominal power rating than turbines in other classes. Classes II and III turbines are intended to produce more energy at lower wind speeds than their Class I counterparts. Class III turbines, in particular, often have a lower high-wind-speed cut-out threshold than the others, meaning the turbines give up generating some energy at higher wind speeds. Many Class III turbines also employ larger rotors and smaller generators relative to their power rating. These differences increase their capacity factor (the average output divided by maximum output), making them more cost-effective at moderate-resource sites.

The turbine class for a wind project is usually determined by considering all the likely locations where turbines might be deployed and finding the lowest appropriate suitability class. If the windiest and most turbulent points in the project area require a Class IIA turbine, then the other turbines will usually be Class IIA as well. Occasionally, a wind project may employ two different turbine models, each corresponding to a different IEC class, to take full advantage of variations in the resource.

In addition to the suitability class, factors to consider in choosing a turbine model include price, warranty and support options, technology maturity and track record, proximity of operating and maintenance facilities, and expected mean output. The analyst might start by listing all available turbines within the site's suitability class. He or she could then contact the manufacturers to obtain pricing, availability, warranty, and other pertinent information, along with a turbine power curve. Using the observed speed frequency distribution from one of the site's monitoring masts, the analyst could then quickly and easily estimate the mean output of each turbine and compare the capital and operating costs per unit of output.

Aside from the hub height, rotor diameter, and rated capacity, the most important turbine characteristics are the power and thrust (force against the wind) produced over a range of wind speeds and air densities. An example of a set of power curves for a range of air densities is shown in Figure 16-5. The software interpolates the power from these data, given the estimated mean air density at the turbine's location and the appropriate speed bin.

If the power curve is available for only one value of air density (such as the standard sea-level density), the output can be estimated by adjusting the speed in proportion to the cube root of the air density, as in the following equation:

$$v_{adj} = v_{site} \left(\frac{\rho_{site}}{\rho_0} \right)^{1/3} \tag{16.1}$$

Wind speed, m/s	Air density, kg/m³											
	1.02	1.04	1.06	1.08	1.10	1.12	1.14	1.16	1.18	1.20	1.22	1.225
0.0	0	0	0	0	0	0	0	0	0	0	0	0
1.0	0	0	0	0	0	0	0	0	0	0	0	0
2.0	0	0	0	0	0	0	0	0	0	0	0	0
3.0	0	0	0	0	0	0	0	0	0	0	0	0
4.0	30	31	32	34	35	36	38	38	40	41	42	42
5.0	94	98	100	103	106	108	111	114	117	119	122	122
6.0	194	199	204	209	214	218	223	228	233	238	242	243
7.0	336	344	351	358	366	374	382	389	397	404	412	414
8.0	526	537	548	559	570	582	593	604	615	626	637	640
9.0	766	782	797	813	828	844	859	874	890	906	921	925
10.0	1065	1087	1109	1131	1154	1176	1198	1220	1242	1265	1287	1293
11.0	1418	1442	1466	1490	1512	1535	1558	1579	1601	1622	1644	1649
12.0	1731	1750	1769	1786	1803	1819	1834	1850	1863	1877	1890	1893
13.0	1926	1937	1946	1954	1960	1966	1970	1974	1978	1980	1981	1982
14.0	1990	1993	1995	1998	1998	1998	1998	1998	1998	1998	1998	1998
15.0	2000	2000	2000	2000	2000	2000	2000	2000	2000	2000	2000	2000
16.0	2000	2000	2000	2000	2000	2000	2000	2000	2000	2000	2000	2000
...	2000	2000	2000	2000	2000	2000	2000	2000	2000	2000	2000	2000
25.0	2000	2000	2000	2000	2000	2000	2000	2000	2000	2000	2000	2000

Figure 16-5. An example of a family of power curves for a range of air densities for a hypothetical 2-MW wind turbine. This curve is defined in 1 m/s increments, although in many cases the power curve is defined in 0.5 m/s increments. In addition, values for air densities above 1.225 kg/m³ are usually provided. *Source:* AWS Truepower.

ρ_{site} is the site air density and ρ_0 is the nominal air density for which the power curve is defined.[1] This adjusted speed is then applied to the power curve as usual. This method is appropriate for pitch-regulated turbines, which are the great majority of large, grid-connected machines. For stall-regulated turbines, which are the predominant

---

[1] The cube root in this equation comes from the cubic relationship between the wind power density and the wind speed (Chapter 10).

type among small wind turbines designed mainly for residential or farm use, the power can be estimated from the following equation:

$$P_{adj} = P \left(\frac{\rho_{site}}{\rho_0} \right) \qquad (16.2)$$

where P is the turbine output for a given speed at the nominal air density. While such adjustments are acceptable for making preliminary energy estimates, it is highly recommended that the analyst obtain a certified power curve for a range of air densities from the manufacturer before finalizing the project's energy production estimate.

16.5 DESIGNING AND OPTIMIZING A TURBINE LAYOUT

After choosing a turbine model, the resource analyst can begin to design the turbine layout. This process usually involves balancing a number of competing goals.

- The first goal is the desire to make maximum use of the land under the developer's control. Up to a point, adding more turbines to a site increases the potential production and, consequently, the revenues. When spread over the fixed capital costs of the project (such as engineering studies, operations and maintenance buildings, and the electrical substation and transmission line), the increased revenues can result in a lower overall cost of energy, making the project more competitive and profitable. Most developers, therefore, want to create as large a project as possible.

- Second, there is the desire to maximize the capacity factor, or average output as a fraction of the rated capacity. Assuming the same installation cost per unit rated capacity, the higher the capacity factor, the greater the profit margin on the project for a given power sales price. This goal usually works against the first, as it implies making use of only the best wind resources in the project area and placing the turbines far enough apart to keep wake losses small.

- The third goal is to minimize the plant installation cost. Aside from the turbines themselves, two key elements of the installation cost are the access roads and electrical collection system connecting the turbines together. By and large, the shorter the required length of roads and cables, the lower the cost. In addition, the steepness of terrain and barriers such as rivers and protected areas must be considered. This goal usually argues for a relatively compact layout with no "stranded" turbines (single turbines or small groups of turbines far from the rest).

- Last, the layout must satisfy the myriad regulatory requirements (such as setbacks from property boundaries), environmental constraints (e.g., noise limits at residences), community concerns (including visual and noise impacts), and other issues that may constrain the placement of turbines. While some of these constraints can be objectively met, others require subjective or qualitative judgments.

The analyst usually starts by manually placing turbines along obvious topographic features or areas of strong resource. If the site has a strongly predominant wind direction, it is a common practice to space the turbines relatively close together, perhaps 3–4 rotor diameters apart, in the prevailing crosswind direction and much farther apart, perhaps 6–10 or more rotor diameters, in the prevailing downwind direction. This practice generally results in close to the maximum number of turbines with manageable wake losses and turbulence-induced loads.

Most software programs come with an optimizing feature that automatically adjusts the turbine locations through hundreds or thousands of iterations to find one that maximizes net energy production (including wake losses) while respecting all setbacks, exclusions, and other objective constraints. These routines are effective, although sometimes very slow for large layouts consisting of hundreds of turbines or more (overnight or weekend runs are common). One drawback is that they do not account for construction costs. The result can be stranded turbines that would be costly to build. It is often necessary to manually adjust the layout by removing or moving such turbines. Some software, such as the openWind Enterprise program, includes a cost-of-energy optimizing feature, which designs road and electrical cabling networks on the fly and takes their cost into account. Using this feature usually results in a more compact layout.

However, no matter how satisfying the optimized solution might be, the developer should expect to have to make adjustments. The GIS data layers used to establish setbacks and other exclusions may not be perfectly accurate. Unmapped features such as buildings and rock outcroppings can make it impossible to install a turbine where planned. Community concerns about noise and appearance may also drive siting decisions. Photo simulations such as that shown in Figure 16-6, which can be produced by most wind plant design software, are often a useful tool for soliciting community feedback and building community support for a project.

16.6 GROSS AND NET ENERGY PRODUCTION

The gross energy production is the annual output of the wind plant without wake or other losses. It is calculated, for each turbine in the layout, from the following equation:

$$E_k = 8766 \sum_{i=1}^{N_d} \sum_{j=1}^{N_v} F_{ijk} P_{ijk} \tag{16.3}$$

The sum is over the number of direction steps N_d and speed bins N_v. F_{ijk} is the frequency of occurrence (expressed as a fraction) and P_{ijk} is the power output for direction sector i, speed bin j, and turbine k. The factor 8766 is the average number of hours in a year (taking into account that one in every 4 years is a leap year, which has 24 extra hours). The units of energy are either kilowatt-hours or megawatt-hours, depending on whether the turbine's power output is expressed in kilowatts or megawatts.

Figure 16-6. Photo simulation of a proposed wind project created by wind plant design software. Such simulations can be useful for soliciting community feedback and building community support. *Source:* AWS Truepower.

The frequencies within the sum are determined by applying speedup factors and directional shifts to the observed speed and direction frequency distributions in the TAB file based on relationships defined in the WRG file. The power output is taken from the power curve interpolated or extrapolated to the appropriate air density. These calculations are all handled automatically by the software, and there is usually little reason for the user to delve into them. However, situations can arise where it becomes helpful to understand them in some depth. For this reason, some of the key methods and equations, along with the important topic of wake modeling, are addressed in Section 16.7.

The net energy production equals the gross production minus losses. A good estimate of plant losses is essential for accurately determining the long-term financial performance of a wind project. A tendency to underestimate losses is one of the reasons why wind plants have often not produced as much energy as predicted in preconstruction studies. In North America, the overestimation averaged around 10% for projects built up to 2008 (1, 2). Improved resource assessment methods, as well as more realistic loss assumptions such as those described below, have largely closed this gap (10).

Table 16-2 provides a breakdown of the main loss categories, each with a range of values (as a percentage of energy production) encountered in practical application. The loss categories are described in more detail below. It should be stressed that every project is different, and some may have higher or lower losses than the ranges shown here. In addition, losses can change over time. Availability losses, in particular, tend

Table 16-2. Loss categories and typical values

Loss category	Low	Typical	High
Wake effects, %	3	6.7	15
Availability, %	2	6.0	10
Electrical, %	2	2.1	3
Turbine performance, %	0	2.5	5
Environmental, %	1	2.6	6
Curtailments, %	0	0	5
Total losses, %	7.8	18.5	37.0

The ranges apply to plants in mature operation; losses, especially availability losses, may be greater in the first 6–12 months after construction. Some plants may have smaller or larger losses than indicated here. The total loss is the product of the efficiencies; see Equation 16.4.
Source: AWS Truepower.

to be greater in the first 6–12 months of operation as problems are identified and resolved. Turbine performance, too, can change over time as turbine components such as blades become worn and pitted.

16.6.1 Wake Effects

This is the reduction in wind speed and increase in turbulence that occurs downstream of a wind turbine. In projects involving more than a handful of turbines, wake effects typically reduce power production by anywhere from 3% to 15%. Keeping this loss manageable is the main reason turbines are rarely spaced closer than 6 rotor diameters in the prevailing wind direction. Furthermore, wake-induced turbulence can cause wear on the components of turbines, and for this reason, turbines are usually spaced no closer than 3 rotor diameters in crosswind directions, and they may have to be shut down under certain conditions to satisfy the manufacturer's warranty.

Since wake effects change with the layout, all wind plant design software must contain a wake model of some kind, and most have more than one. Three main types in regular use are described later in this chapter: the Park and modified Park model, the eddy viscosity (EV) model, and the deep-array (or large-array) wake model. Other models based on CFD and LES (large-eddy simulation) codes, which are under development, are outside the scope of this book.

16.6.2 Downtime

A wind plant or turbine is said to be available when it is capable of generating its full rated output, given sufficient wind. Availability losses occur when some turbines in a project, or the entire project, are inoperative for some reason. They can also occur because of a failure or shutdown of the power grid or substation. An overall plant availability of 97–98% (2–3% loss) is frequently assumed in energy production studies, but is likely to be optimistic unless there is good evidence that the plant operator has regularly achieved such high performance with the turbine model in question. Plant start-up problems, repair delays, fleet-wide turbine issues requiring retrofits, and other issues can cause extended periods of downtime that reduce the lifetime average availability. Some observers have also noted a tendency for turbines to break down under high wind conditions, which increases the energy loss (e.g., 3% downtime may actually correspond to a 5% energy loss). An average availability loss of 2–10% is typically encountered in operation.

16.6.3   Electrical Losses

Losses are experienced in all electrical components of the wind project, including the padmount transformer, electrical collection system, and substation transformer. These losses are established in the electrical system design. A loss range of 2–3% is typical.

16.6.4   Turbine Performance

This factor includes the effects of suboptimal turbine control settings (e.g., yaw misalignments, control anemometer calibration errors, and blade pitch inaccuracies or misalignments), high wind control hysteresis (accounting for the fact that a turbine which has been shut down because of high winds must wait until the speed drops below a lower speed threshold before restarting), and high turbulence, shear, or inclined flow departing from the conditions for which the power curve has been defined. These losses individually tend to be small, but at sites experiencing unusually high winds, high turbulence, or other extreme conditions, they can be as much as 3% in the aggregate. In addition, there is evidence that turbines often fall short of their advertised power curves even in IEC-compliant power curve tests (10). This may justify an additional 2–3% loss.

16.6.5   Environmental Losses

This category includes losses due to the accumulation of ice on the blades, blade soiling and degradation, shutdowns triggered by very high or very low temperatures or by lightning, and the difficulty of accessing sites in bad weather to carry out repairs. These losses may be estimated in some cases based on information collected from the site, such as temperature records and the frequency of icing observed in the wind data validation, or from regional records such as lightning frequency maps and snowfall records. Remote sites that are likely to be inaccessible in bad weather, including offshore sites, often incur greater than normal losses.

16.6.6   Curtailments

If turbines are spaced closer than 3 rotor diameters from each other, the manufacturer may impose a curtailment strategy to limit wear and tear caused by wake-induced turbulence. This typically requires that some turbines (such as every other turbine in a row) be shut down when the wind is from a certain range of directions and above a certain speed threshold.

In addition, the utility company or grid operator can impose plant-wide curtailment as a means to help manage the transmission grid. As wind power's penetration on utility systems has grown, plant-level curtailments have become more common, and in some regions, their impact is much greater than the "high" estimate cited here. Curtailment may also be imposed at certain times for environmental reasons, such as to avoid interfering with bird and bat movements, satisfy nighttime noise restrictions, and reduce shadow flicker (caused by sunlight reflecting from blades).

Combining Losses. The individual loss factors are combined not by direct addition but by multiplying the efficiencies (defined as one minus the loss). The total loss is given by,

$$L_{\text{total}} = 100\% - (100\% - L_1)(100\% - L_2)(100\% - L_3)\ldots \qquad (16.4)$$

where the values L_n are the individual losses in percent (only the first three being shown in this equation).

16.7 SPECIAL TOPICS

Although one can run wind plant design software without understanding the details of the underlying calculations, it is useful to know something about how the programs work, as this can help the user make sensible choices and diagnose problems. The following sections discuss methods of extrapolating the wind resource from one or more masts to the turbine locations, the gross energy calculation, and wake modeling.

16.7.1 Extrapolating the Wind Resource from Mast to Turbine

Although the details differ, all the leading programs make use of a concept called the *speedup ratio*, which is the ratio of the speed predicted at a point to the speed at a mast for a particular direction. This ratio is calculated using the information contained in the WRG and point WRG files. The speedup ratio for each direction is multiplied by the observed speed at each mast to estimate the corresponding speed at the point.

$$v_{pi} = R_{pmi} v_{mi} \qquad (16.5)$$

v_{pi} is the predicted speed at point p for direction i, v_{mi} is the corresponding speed at the mast m, and R_{pmi}, is the speedup ratio from the mast to the point for the same direction.

The speedup ratio for a given direction is generally assumed to be constant with speed. Thus, the observed speed frequency distribution at the mast is merely scaled up or down by the same factor but is not otherwise changed. In fact, of course, the speed frequency distribution *can* vary from point to point. Some programs use the predicted variation in the Weibull k parameter in the WRG file in an attempt to estimate changes in the observed frequency distribution. However, there is little hard data to confirm whether this option improves accuracy or not.

The method of directional speedups also generally assumes that the directional frequencies do not vary across the project area. Within the software, this assumption is realized by setting the directional frequencies at all points equal to the mast frequencies. Although this approach works reasonably well in many cases, it can be problematic where winds are heavily influenced by the terrain and land surface properties. For example, winds channeled through a mountain pass may exhibit very different directions at different points within and outside the pass. Similarly, as winds transition from offshore to onshore, and vice versa, there is frequently a change in direction caused by the contrasting thermal properties and roughness of the land and ocean surfaces.

To handle such situations, most programs include a "directional shift" or similar option, which makes a guess, based on the modeled variation in directional frequencies, about how the observed frequencies differ between the mast and a given point. Again, there is little evidence to say whether this adjustment improves accuracy, or by how much, and its effect is likely to be very site-dependent.

16.7.2   Using Multiple Masts

As noted in Chapter 13, most wind energy projects employ more than one mast. This poses the practical challenge of combining the information from the various masts in estimating the energy production.

One common approach is to divide the project area into sections, each of which is assigned to one mast. The sections may be defined by distance (i.e., the closest mast is assumed to "dominate" the area), or they may be defined by some other criterion, such as topographic similarity (e.g., ridgetop sections are assigned to ridgetop masts). The simulation for each mast is done separately from the others, and the wind resource for any turbines within a section is extrapolated from whichever mast that section is assigned to.

This approach is pragmatic, but it can be awkward when, as often happens, there is a discontinuity in the predicted wind resource where two sections meet. The resulting energy production estimate can change abruptly (and unrealistically) when a turbine is moved from one side of the dividing line to the other. This problem is a reflection of the inaccuracies of wind flow modeling, for if the models were perfect, the predicted wind resource at a point would be the same (within the measurement uncertainty at the masts) no matter which mast was used to initialize or adjust the model.

A more esthetically pleasing, if not necessarily more accurate, approach is to smoothly blend the predicted wind resource from the different masts. This method adopts the assumption that every mast offers at least some useful information about the wind resource at any point, and that the weighted average of several estimates should be more reliable than any single estimate alone. The challenge is to determine a suitable method of weighting. A relatively simple blending technique is to weight each mast's prediction according to the inverse of the squared distance to that mast. Extending Equation 16.1,

$$\overline{v_{pij}} = \frac{\displaystyle\sum_{j=1}^{M} \frac{v_{pij}}{d_{pj}^2 + C}}{\displaystyle\sum_{j=1}^{M} \frac{1}{d_{pj}^2 + C}} \tag{16.6}$$

Here, the sum is over M, the number of masts, v_{pij} is the predicted speed at point p for direction i based on mast j, d_{pj} is the distance between the point and the same mast, and C is a smoothing constant that prevents the equation from becoming undefined very close to a mast. Since most wind flow models do not perform this type of blending, it is usually necessary to write a software to do it. Another approach used by some analysts is to estimate the energy production for each turbine using one mast at a time and then to blend the estimates "offline" in a spreadsheet program.

Distance-weighted blending is relatively easy, but is it the best approach? Not necessarily. It assumes implicitly that the uncertainty associated with the prediction from any given mast depends strictly on the distance to that mast.[2] As noted in Chapter 13, however, distance is only one factor influencing the accuracy of wind flow modeling. The modeling uncertainty and, therefore, the appropriate blending weight also depend on the similarity of topographic and other conditions between the point and the mast. For example, if the point were on a ridgetop, it might be better to give a ridgetop mast more weight in the adjustment than a mast down the slope, even if the latter were closer. Although an approach based on topographic similarity and other factors is more defensible, it is also more difficult to put into practice, as it requires an understanding of how the modeling uncertainty varies with these factors (3).

16.7.3 Wake Modeling

Wake modeling remains an area of active research because of the great complexity and wide range of scales of turbine–atmosphere interactions. While the basic physical equations are well understood, a complete numerical solution to the wake problem remains beyond the capability of today's computers.

Two early-generation models, the Park and EV models, are currently in wide use, and a third class of model, the so-called deep-array models designed for large projects, has recently appeared.

Park and Modified Park Models. The Park model was developed in the mid 1980s (4) and has been implemented in the WAsP software as well as in most wind plant design programs. It characterizes a turbine wake by two parameters: the width and the speed deficit relative to the free-stream speed. The width D is assumed to be initially equal to the rotor diameter and to grow linearly with distance downstream:

$$D(x) = D_0(1 + 2kx) \qquad (16.7)$$

---

[2]The connection between distance and uncertainty is implied by statistical theory, which holds that independent measurements of the same quantity should be combined in a weighted average, where the weight accorded each measurement is inversely proportional to its uncertainty squared. It can be shown that the new estimate produced this way has the lowest possible uncertainty. Substituting distance for uncertainty yields the standard distance-weighted blending method.

Here D_0 is the rotor diameter (in meters), k is the decay constant, and x is the distance downstream from the rotor (expressed in rotor diameters). The speed deficit δv (as a fraction of the free-stream speed) is assumed to be constant across the width and is given by the equation,

$$\delta v(x) = (1 - \sqrt{1 - C_t}) \left(\frac{D_0}{D(x)} \right)^2 \qquad (16.8)$$

In this equation, C_t is the thrust coefficient of the turbine, a parameter that represents the amount of force generated by the rotor against the wind. Along with the turbine power curve, C_t is specified by the manufacturer and varies with speed.

A key question in wake modeling is how the wakes of successive downstream turbines combine with one another. In the Park model, the incident wind speed at a downwind turbine is the free-stream wind speed minus the wake deficit calculated by Equation 16.8, multiplied by the fractional overlap between downwind turbine's rotor and the upwind turbine's wake cross section (Fig. 16-7). Where there are several upstream wind turbines with overlapping wakes, the incident speed is assumed to equal the free-stream speed minus the largest single wake deficit.

Modified Park differs from Park in two ways. One is how the overlap with an upwind wake is calculated: in modified Park, the area of overlap is represented by a rectangle. The other is that the wake generated by a wake-affected turbine is the same as if the turbine were in the free-stream except that its thrust coefficient, C_t, is calculated using the incident wind speed for that turbine. In general, these modifications produce a smaller wake loss than that estimated by Park. Where there is a choice, modified Park is generally used.

The only parameter that needs to be defined when running the Park or modified Park model is the wake decay constant. This typically ranges in value from 0.04

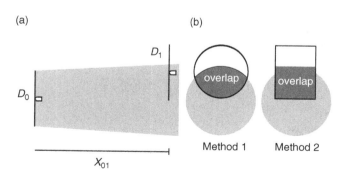

Figure 16-7. (a) Plan view of a turbine wake of initial width D_0 expanding with distance downstream and intersecting the rotor of a second turbine. (b) Two methods of calculating the area of overlap between the wake and the second turbine. The first is that implemented in the Park model, the second is that implemented in the modified Park model. *Source:* AWS Truepower.

to 0.075. Lower values produce longer-lived wakes such as might be seen offshore. Higher values produce wakes that decay faster.

Eddy Viscosity (EV) Model. The EV model was developed in the late 1980s around the same time as the Park model (5). Unlike Park, which adopts a purely empirical approach, the EV model is a kind of CFD model that solves the following simplified form of the Navier–Stokes equations:

$$U\frac{\delta U}{\delta x} + V\frac{\delta U}{\delta r} = \frac{\varepsilon}{r}\frac{\delta (r\delta U/\delta r)}{\delta r} \tag{16.9}$$

The equation is in cylindrical coordinates: r is the radius from the center line of the rotor and x is the distance downstream. U is the speed in the downwind direction, and V is the radial speed. The parameter ε is the eddy viscosity, which represents the friction exerted by adjacent turbulent eddies.

The initial wake-induced wind speed deficit is assumed to be a Gaussian (bell-shaped) curve, which starts 2 rotor diameters downstream of the turbine. At the center, the speed deficit (again, as a fraction of the free-stream speed) is

$$\delta v_c = C_t - 0.05 - [16C_t - 0.5]\frac{I_0}{1000} \tag{16.10}$$

I_0 is the ambient turbulence intensity. The shape of the deficit curve is given by the Gaussian equation,

$$\delta v(r) = \delta v_c e^{-\frac{r^2}{w^2}} \tag{16.11}$$

where the effective wake width w is defined by

$$w = R\sqrt{\frac{C_t}{8\delta v_c(1 - 0.5\delta v_c)}} \tag{16.12}$$

R is the rotor radius.

Once initialized, the wake propagates downstream, expanding and dissipating as the air within it mixes with the surrounding free-stream air. The rate of mixing is determined by the eddy viscosity, which is a function of the ambient turbulence—the greater the turbulence, the greater the rate of mixing and the faster the wake deficit decays. Thus, the EV model contains no empirical wake decay constant: the decay rate is determined by the model equations, and the only inputs, aside from the characteristics of the turbine, are the ambient speed and turbulence intensity.

Since Equation 16.9 has no analytical solution, it is solved numerically using a finite differencing technique, a standard CFD method.

Deep-Array Wake Models. In the past several years, researchers have become aware that the current generation of wake models may underestimate wake losses in large wind projects with multiple rows of wind turbines. The crux of the problem appears to be that the leading wake models, including the Park, modified Park, and EV models, ignore two-way interactions between the atmosphere and the turbines (6). Each turbine extracts energy from the wind passing through its rotor plane, creating a zone of reduced speed extending some distance downstream. Upstream and outside this zone of influence, it is assumed the ambient wind is unaffected.

Both theory and experiment suggest that for large arrays of wind turbines, this assumption does not hold. The presence of numerous large wind turbines in a limited area can alter the wind profile in the planetary boundary layer (PBL), both within and around the array, thereby reducing the amount of energy available to the turbines for power production. Experimental data supporting this hypothesis comes mainly from offshore wind projects, where the contrast between the drag induced by the turbines and the relatively low roughness of the ocean surface makes the so-called deep-array effect especially pronounced. Onshore, the effect is attenuated, but theory suggests it may nonetheless be significant in large projects.

It has become clear that new models are required that can simulate deep-array wake effects with reasonably good accuracy. Predicting the overall impact of a large wind turbine array is a complex problem involving dynamic interactions between the turbines and various properties of the atmosphere, including vertical and horizontal gradients of temperature, pressure, and speed, as well as turbulence. This problem can be solved completely only through sophisticated numerical modeling requiring very fast computers. However, it may be hoped that simplified approaches will work well enough for wind projects likely to be developed in the next several years.

The most widely implemented deep-array wake model is based on a theory advanced by Sten Frandsen (7), in which an infinite array of wind turbines is represented as a region of uniform high surface roughness. The roughness imposes drag on the atmosphere, causing a downstream change in the structure of the PBL and, in particular, a reduction in the free-stream wind speed at the turbine hub height. According to this theory, the wind farm equivalent roughness z_{00} is given by

$$z_{00} = h_{\mathrm{h}} \exp\left(-\frac{\kappa}{\sqrt{c_{\mathrm{t}} + \left(\kappa/\ln(h_{\mathrm{h}}/z_0)\right)^2}}\right) \qquad (16.13)$$

In this equation, h_{h} is the hub height, κ is the von Karman constant (about 0.4), z_0 is the background roughness between turbines, and c_{t} is the distributed thrust coefficient, defined as

$$c_{\mathrm{t}} = \frac{\pi}{8 s_{\mathrm{d}} s_{\mathrm{c}}} C_{\mathrm{t}}$$

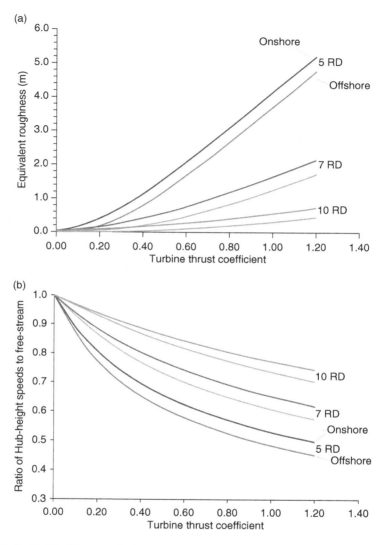

<u>Figure 16-8.</u> (a) Wind farm equivalent roughness z_{00} for an infinite turbine array as a function of thrust coefficient and array spacing in rotor diameters (RDs), for both offshore (ambient $z_0 = 0.001$ m) and onshore (ambient $z_0 = 0.03$ m) projects. (b) Asymptotic hub-height wind speed as a fraction of the free-stream speed for the same cases. *Source:* AWS Truepower, based on equations in Frandsen (2007).

where C_t is the turbine thrust coefficient and s_d and s_c are the mean downwind and crosswind spacings, respectively, in rotor diameters. Figure 16-8a shows z_{00} for a range of C_t and mean array spacings $(s_d s_c)^{0.5}$. The roughness is strongly dependent on the spacing, much less so on the background or ambient roughness (here assumed to be 0.001 m for offshore arrays and 0.03 m for onshore arrays).

Once the equivalent roughness is defined, the impact on the hub-height wind speed deep within the array (i.e., where the PBL has reached equilibrium with the array roughness) is estimated from meteorological theory under the assumption of a constant geostrophic wind speed G and a neutral logarithmic profile throughout the PBL. The result is approximated by the following equation.

$$\frac{v_h'}{v_h} = \left(\frac{z_{00}}{z_0}\right)^{0.07} \frac{\ln\left(h_h/z_{00}\right)}{\ln\left(h_h/z_0\right)} \tag{16.14}$$

Here, v_h' and v_h are the hub-height speeds deep within the array and far upstream, respectively. The results are plotted in Figure 16-8b.

An important issue with the Frandsen theory is that it does not address the wake effects of individual turbines. Instead, it treats the array as an infinite sea of undifferentiated surface drag. This means that the predicted wind resource at a particular location does not depend on whether there are any turbines immediately upwind or not, which is unrealistic. Furthermore, the roughness depends on the array density, which implies that in practical application it would have to be recalculated every time the layout is modified. Thus, to be useful for wind project design and optimization, the Frandsen theory must either be modified or combined in some way with other methods.

One solution that has been offered to this problem is to model turbines as discrete roughness elements. Each turbine is surrounded by an area where the surface roughness is increased. When the free-stream wind passes the front edge of this area, an IBL is created which grows with distance downstream, and within which the wind shear is increased. When the wind passes the rear edge of the rough area, another IBL is created which grows at a different (generally slower) rate, and within which the shear reverts to the normal, background value. The overall effect is to reduce the wind speed at heights below the top of the first IBL.

The effect is illustrated in Figure 16-9, where the two IBLs cast by the first turbine are shown. The top curve represents the IBL created by the high roughness of the turbine, and the bottom curve represents the IBL created by the transition back to the low ambient roughness between turbines. The two curves to the right show the disturbed and free-stream wind speed profiles; the difference between them is the wind speed deficit.

This general approach has been implemented in several programs including Wind-Farmer and openWind. Published reports indicate that they are capable of matching observed "deep-array" wake effects in offshore projects with reasonable accuracy. Evidence from onshore projects is more in dispute (8, 9). Regardless, the amount of data available to validate the models is exceedingly small, so it is safe to say that they remain developmental, and it is unknown how well they will work at much larger projects, either onshore or offshore. For the time being, caution argues for using one of the available deep-array wake models in analyzing large projects with at least four rows of turbines.

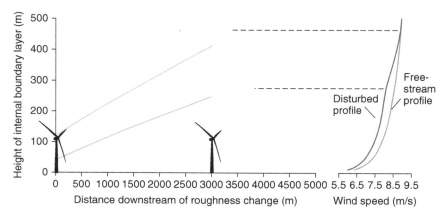

<u>Figure 16-9.</u> Illustration of the growth of two IBLs created by the wind turbine at left, which is modeled as a surface roughness change. Between the top and bottom curves, the wind shear is determined by the high roughness of the turbine; above the top curve and below the bottom curve, it reverts to the shear associated with the background roughness. The two curves on the right illustrate the effect of this turbine on the free-stream wind speed profile at the fourth turbine, 3000 m directly downwind of the first. Additional turbines (not shown) would create their own IBLs and further modify the profile in an analogous fashion. The curves are examples only and are not intended to represent any particular model or plant. *Source:* AWS Truepower.

16.8   QUESTIONS FOR DISCUSSION

1. Suppose the windiest point in a wind project site has a mean wind speed of 8.6 m/s and an estimated annual average air density of 1.125 kg/m³. The mean turbulence intensity at 15 m/s is 15%. What turbine suitability class would be appropriate for this site? Using the Internet, find three commercial megawatt-class turbines in this class and list their main characteristics (rated capacity, rotor diameter, hub height).

2. Consider the topographic map in Figure 16-2. Identify the stranded turbines in this layout, and explain why they are likely to be more costly to build than other turbines in the layout.

3. Consider the set of power curves in Figure 16-5. Suppose you are only given the power curve for an air density of 1.225 kg/m³ (last column on the right). Assume that the estimated air density at your site is 1.14 kg/m³. Using Equation 16.1, calculate a new power curve for the site, and then compare it with the true power curve for the same air density from the table. What is the maximum difference at any given speed? Do you think this is a reasonable approximation? (Procedure for calculating the power curve: adjust the standard speeds for the existing power curve according to Equation 16.1, interpolate the power at 1.225 kg/m³ to the adjusted speeds, and assign the new power values to the original, unadjusted speeds.)

4. Suppose a wind project has the following losses: wakes, 7.2%; availability, 4.6%; electrical, 2.5%; and other, 2%. What is the total loss?

5. Using Equation 16.3, and with the speed frequency distribution in the table below, the power curves in Figure 16-5, and assuming an air density of 1.20 kg/m^3, estimate the annual average gross power output in gigawatt-hours of a wind project consisting of 20 turbines (40 MW). With the "typical" losses shown in Table 16-2, estimate the net annual power output. What are the expected annual average gross and net capacity factors for this turbine?

Speed (m/s) (top of bin)	Frequency (%)
1	0.06%
2	1.06%
3	4.06%
4	10.09%
5	17.29%
6	18.42%
7	14.35%
8	11.92%
9	9.66%
10	5.92%
11	3.44%
12	2.34%
13	1.65%
14	1.09%
15	0.75%
16	0.51%
17	0.30%
18	0.19%
19	0.13%
20	0.07%
21	0.04%
22	0.03%
23	0.02%
24	0.01%
25	0.00%

6. Download and install the open-source (Community) version of the openWind program (www.awsopenwind.org), or the demonstration version of another wind plant design program, if one is available. Download and step through the tutorial.

REFERENCES

1. White E. Closing the gap on plant underperformance: A Review and calibration of AWS Truewind's energy estimation methods," USA: AWS Truepower: July 2009.

2. Johnson C, Tindal A, Harman K, Graves A, Hassan G, Validation of energy predictions by comparison to actual production. In: AWEA Windpower 2008 Conference, June 2008, Houston, Texas.

3. Brower MC, Robinson NM, Hale E. Wind flow modeling uncertainty: quantification and application to monitoring strategies and project design. USA: AWS Truepower; 2010. Available at http://www.awsopenwind.org/downloads/documentation/ModelingUncertaintyPublic. pdf. (Accessed 2012).

4. Katic I, Højstrup J, Jensen NO. A simple model for cluster efficiency. In: Proceedings of European Wind Energy Conference and Exhibition; Rome; 1986. pp. 407–410.

5. Ainslie JF. Calculating the flowfield in the wake of wind turbines. J Wind Eng Ind Aerodyn 1988;27:213–224.

6. Frandsen ST, Barthelmie RJ, Pryor SC, Rathmann O, Larsen S, Højstrup J, Thøgersen M. Analytical modeling of wind speed deficit in large offshore wind farms. Wind Energy 2006;9:39–53.

7. Frandsen ST. Turbulence and turbulence-generated structural loading in wind turbine clusters. Report Risø-R-1188(EN). Risø National Laboratory; Jan 2007.

8. Wolfe J. Deep array wake loss in large onshore wind farms (a model validation). Oklahoma City, Oklahoma, USA: AWEA Wind Resource Assessment Workshop 2010. Available at http://www.ramwind.com/ publications/20100818.AWEA.DeepArray.Wolfe.pdf. (Accessed 2012).

9. Brower MC, Robinson NM. The openWind deep-array wake model: development and validation. USA: AWS Truepower; 2010. Available at http://www.awsopenwind.org/downloads/ documentation/DAWM_WhitePaper.pdf. (Accessed 2012).

10. Bernadett DW, Backcast KB. Verifying the accuracy of energy and uncertainty estimates. USA: AWS Truepower; May 2012.

SUGGESTIONS FOR FURTHER READING

Ainslie JF. Calculating the flowfield in the wake of wind turbines. J Wind Eng Ind Aerodyn 1988;27:213–224.

Frandsen ST, Barthelmie RJ, Pryor SC, Rathmann O, Larsen S, Højstrup J, Thøgersen M. Analytical modeling of wind speed deficit in large offshore wind farms. Wind Energy 2006;9:39–53.

International Electrotechnical Commission (IEC) 61400-1, Wind turbines—Part 1: Design Requirements (Third Edition: 2005–08). (IEC publications can be purchased or downloaded from http://webstore.iec.ch/.)

Katic I, Højstrup J, Jensen NO. A simple model for cluster efficiency. In: Proceedings of European Wind Energy Conference and Exhibition; Rome; 1986. pp. 407–410.

Robinson NM. openWind theoretical basis and validation, AWS Truepower; 2010. Available at http://www.awsopenwind.org/downloads/documentation/OpenWindTheoryAndValidation. pdf. (Accessed 2012).

Wind Farm Design: Planning, Research, and Commissioning. Renewable Energy World; April 2009. Available at http://www.renewableenergyworld.com/rea/news/article/2009/04/wind-farm-design-planning-research-and-commissioning. (Accessed 2012).

Wind Turbine Wake Aerodynamics. Progress in Aerospace Sciences (39), Vermeer, Sorensen, and Crespo; 2003. Available at http://citeseerx.ist.psu.edu/viewdoc/download?doi=10.1.1.132.6485&rep=rep1&type=pdf. (Accessed 2012).

WIND RESOURCE ASSESSMENT EQUIPMENT VENDORS

INSTRUMENT AND TOWER SUPPLIERS

All Weather, Inc.
www.allweatherinc.com

Belfort Instrument
www.digiwx.com

Campbell Scientific, Inc.
www.campbellsci.com

Climatronics Corporation
www.climatronics.com

Coastal Environmental Systems
www.coastalenvironmental.com

Geotech Instruments
www.geoinstr.com

Kipp & Zonen
www.kippzonen.com

LI-COR, Inc.
www.licor.com

Met One Instruments
www.metone.com

NovaLynx Corporation
www.novalynx.com

NRG Systems, Inc.
www.nrgsystems.com

Radian Corporation
www.radiancorp.com

M. Young
www.youngusa.com

Rohn Products
www.rohnnet.com

Sabre Industries
www.sabreindustriesinc.com

Scientific Sales
www.scientificsales.com

Second Wind Inc.
www.secondwind.com

Thies Clima
www.ThiesClima.com

Tower Systems
www.towersystems.com

Vaisala Inc.
www.vaisala.com

Vector Instruments
www.windspeed.co.uk

Yankee Environmental Systems
www.yesinc.com

WindSensor
www.windsensor.dk

LIDAR EQUIPMENT SUPPLIERS

Catch the Wind Ltd.
www.catchthewindinc.com

Leosphere
www.lidarwindtechnologies.com

NRG Systems
www.nrgsystems.com

Lockheed Martin
www.lockheedmartin.com

Natural Power
www.naturalpower.com

Wind Resource Assessment: A Practical Guide to Developing a Wind Project, First Edition.
Michael Brower et al.
© 2012 John Wiley & Sons, Inc. Published 2012 by John Wiley & Sons, Inc.

SgurrEnergy Ltd.
  www.sgurrenergy.com
**SODAR EQUIPMENT
  SUPPLIERS**
**Atmospheric Research &
  Technology (ART)**
www.sodar.com

**Atmospheric Systems
  Corporation (ASC)**
www.minisodar.com
AQSystem
  www.aqs.se
Metek GmbH
  www.metek.de

Remtech
  www.remtechinc.com

Scintec Corporation
  www.scintec.com

Second Wind Inc.
  www.secondwind.com

WIND RESOURCE ASSESSMENT EQUIPMENT

This a list of some of the most common wind resource assessment equipment currently used within the industry. It should not be regarded as comprehensive.

Cup Anemometers
- Climatronics: F460 Wind Speed Sensor
- Met One: 010C
- Second Wind: C3
- NRG Systems: Maximum #40
- Thies: First Class Advanced
- Vaisala: WA15
- Vector: A100LK
- WindSensor: P2546A

Wind Vanes
- Climatronics: F460 Wind Direction Sensor
- Met One: 020C
- NRG Systems: 200P
- Thies Clima: First Class
- Vaisala: WA15
- Vector: W200P

Vertical Prop Anemometers
- Climatronics: M102236 Vertical Propeller Anemometer
- R. M. Young: 27106 Vertical Propeller Anemometer

Sonic Anemometers
- Applied Technologies: CATI/2
- Campbell Scientific: CSAT3
- Climatronics: 102642 Sonic Wind Sensor
- Gill Instruments: WindSonic 2-D Ultrasonic Anemometer
- Met One: Model 50.5 Solid State Wind Sensor

- Metek: Ultrasonic Anemometer USA-1
- R. M. Young: Model 81000 Ultrasonic Anemometer
- Thies Clima: Ultrasonic Anemometer 3D
- Vaisala: WS425

SODAR Units

- ART: Model VT-1 SODAR System
- ASC: Models 4000, 3000, and 2000
- Metek: Phased Array SODAR
- Remtech: PA0
- Second Wind: TRITON Sonic Wind Profiler
- Scintec Corporation: Flat Array Sodar Acoustic Profiler

LIDAR Units

- Catch the Wind: Vindicator
- Lockheed Martin: WindTracer
- Natural Power: ZephIR Laser Anemometer
- NRG Systems/Leosphere: Windcube
- SgurrEnergy: Galion

SELECTED SOURCES OF GIS DATA

SOURCES OF GIS DATA RELATED TO WIND RESOURCE ASSESSMENT

Region and Data Type	Web Address
UNITED STATES	
FEDERAL	
USGS Seamless Data Server	http://seamless.usgs.gov/index.php
STATE	
State GIS Clearinghouses	http://web.mit.edu/dtfg/www/data/data_gis_us_state.htm
State & Federal GIS Clearinghouses	http://ncl.sbs.ohio-state.edu/5_sdata.html

---

Wind Resource Assessment: A Practical Guide to Developing a Wind Project, First Edition.
Michael Brower et al.
© 2012 John Wiley & Sons, Inc. Published 2012 by John Wiley & Sons, Inc.

Region and Data Type	Web Address
SPECIFIC GIS DATA SITES	
National Wetlands Inventory	http://www.fws.gov/wetlands/Data/mapper.html
Free Orthoimagery Sources	http://worldwindcentral.com/wiki/Sources_of_free_orthoimagery
Great Lake Information Network	http://gis.glin.net/ogc/services.php?by=topic
FCC Data	http://wireless.fcc.gov/geographic/index.htm?job=home
WIND RESOURCE DATA	
Local Universities	Various
NOAA's National Climatic Data Left	http://gis.ncdc.noaa.gov/geoportal/catalog/main/home.page
NREL Wind Power Data	http://www.nrel.gov/gis
State Climate Offices	http://www.stateclimate.org
Wind Powering America	http://www.windpoweringamerica.gov/wind_maps.asp
GLOBAL	
GIS Data Clearinghouse	http://data.geocomm.com
Global Landcover Data	http://www.mdafederal.com/home
Corine Landcover Data (Europe)	http://www.eea.europa.eu/data-and-maps/data/corine-land-cover-2006-raster-1
CDED (Canadian Elevation Data)	http://www.geobase.ca/geobase/en/data/cded/index.html
ASTER Global Elevation Data	http://www.gdem.aster.ersdac.or.jp/index.jsp
SRTM Global Elevation Data	http://www2.jpl.nasa.gov/srtm/
WIND RESOURCE DATA	
AWS Truepower	http://windnavigator.com
3Tier	http://www.3tiergroup.com
Vortex	http://www.vortex.es

Wind Resource Assessment: A Practical Guide to Developing a Wind Project, First Edition.
Michael Brower et al.
© 2012 John Wiley & Sons, Inc. Published 2012 by John Wiley & Sons, Inc.